Housing and Home Unbound

T0173838

Housing and Home Unbound pioneers understandings of housing and home as a meeting ground in which intensive practices, materials and meanings tangle with extensive economic, environmental and political worlds. Cutting across disciplines, the book opens up the conceptual and empirical study of housing and home by exploring the coproduction of the concrete and abstract, the intimate and institutional, the personal and collective.

Exploring diverse examples in Australia and New Zealand, contributors address the interleaving of money and materials in the digital commodity of real estate, the neoliberal invention of housing as a liquid asset and source of welfare provision, and the bundling of car and home in housing markets. The more-than-human relations of housing and home are articulated through the role of suburban nature in the making of Australian modernity, the marketing of nature in waterfront urban renewal, the role of domestic territory in subversive social movements such as seasteading and tiny houses, and the search for home comfort through low-cost energy-efficiency practices. The transformative politics of housing and home are explored through the decolonising of housing tenure, the shaping of housing policy by urban social movements, the lived importance of marginal spaces in Indigenous and other housing, and the affective lessons of the ruin. Beginning with the diverse elements gathered together in housing and home, the text opens up the complex realities and possibilities of human dwelling.

Nicole Cook, University of Melbourne, Australia.

Aidan Davison, University of Tasmania, Australia.

Louise Crabtree, University of Western Sydney, Australia.

'Home is where they have to take you in, to paraphrase Robert Frost. In that inclusive spirit, Nicole Cook, Aidan Davison and Louise Crabtree open the Australian home to a rich and lively group of invitees, who rarely come together under the same scholarly roof. Straddling housing and home, the volume carefully moves between the affective and the abstract, the domestic and the institutional, and the particular and the collective, opening promising lines of new enquiry.'

Nicholas Blomley, Professor of Geography,
Simon Fraser University, Canada

Housing and Home Unbound

Intersections in economics, environment and politics in Australia

Edited by Nicole Cook,
Aidan Davison and Louise Crabtree

Taylor & Francis Group

LONDON AND NEW YORK

First published 2016 by Routledge

2 Park Square, Milton Park, Abingdon, Oxon, OX14 4RN

605 Third Avenue, New York, NY 10017

*Routledge is an imprint of the Taylor & Francis Group,
an informa business*

First issued in paperback 2020

British Library Cataloguing in Publication Data
A catalogue record for this book is available from the British Library

Library of Congress Cataloging-in-Publication Data
Names: Cook, Nicole Therese, editor. | Davison, Aidan, 1966- editor. |
Crabtree, Louise, editor.
Title: Housing and home unbound : intersections in economics,
environment and politics in Australia / edited by Nicole Cook, Aidan
Davison and Louise Crabtree.
Description: Abingdon, Oxon ; New York, NY : Routledge, 2016. |
Includes bibliographical references and index.
Identifiers: LCCN 2015044284 | ISBN 9781138948976 (hardback :
alk. paper) | ISBN 9781315669342 (ebook)
Subjects: LCSH: Housing--Australia. | Housing development--
Australia. | Housing policy--Australia. | Dwellings--Australia.
Classification: LCC HD7379.A3 H56 2016 | DDC 333.33/80994--dc23
LC record available at http://lccn.loc.gov/2015044284

ISBN: 978-1-138-94897-6 (hbk)
ISBN: 978-0-367-73716-0 (pbk)

Typeset in Sabon
by Saxon Graphics Ltd, Derby

Contents

Housing/home and worlds of nature   93

  6 Secure in the privacy of your own nature: political ontology,
     urban nature and home ownership in Australia   99
     AIDAN DAVISON

  7 Making nature and money in the East Perth redevelopment   116
     LAURENCE TROY

  8 Displacement as method: seasteading, tiny houses and
     'Freemen on the Land'   134
     LORENZO VERACINI

  9 'The best house possible': the everyday practices and
     micro-politics of achieving comfort in a low-cost home   151
     MICHELLE GABRIEL, MILLIE ROONEY AND PHILLIPA WATSON

PART III
Housing/home and worlds of possibility   167

 10 Unbounding home ownership in Australia   173
     LOUISE CRABTREE

 11 Performing housing affordability: the case of Sydney's
     green bans   190
     NICOLE COOK

 12 Interstitial housing space: no centre just borders   204
     WENDY STEELE AND CATHY KEYS

 13 Burnt houses and the haunted home: reconfiguring the
     ruin in Australia   218
     KATRINA SCHLUNKE

     Thrown-togetherness, 2015: exegesis   232
     ANDREW GORMAN-MURRAY

     Index   236

Figures

PART III: HOUSING/HOME AND WORLDS OF POSSIBILITY

Acknowledgements

This edited collection emerged from a session at the 2014 Institute of Australian Geographers/New Zealand Geographical Society joint conference at the University of Melbourne, titled *(Un)bounding Housing and Home*. We very much thank the 12 contributors to that exciting session. We are grateful to Professors Susan J. Smith and Keith Jacobs for their encouraging, generous and thoughtful feedback on a draft book proposal, which served to sharpen the book's purpose and our thinking. Thanks are due to all of our contributors for their excellent chapters and collegiality, and to Andrew Gorman-Murray for his evocative, rich and enticing visual contribution to the book project. Lastly, thanks are due to the team at Routledge for the invitation to convert the conference session into this collection, and their encouragement and expert support throughout the process.

Contributors

Fiona Allon is Australian Research Council (ARC) Future Fellow and Senior Lecturer in the Department of Gender and Cultural Studies at the University of Sydney, Australia. She is the author of *Renovation Nation: Our Obsession with Home* (2008). Her recent research focuses on the everyday life of households, specifically in relation to the interface between home/housing, mortgage and financial markets. Parts of this research have been published in the *Journal of Cultural Economy*, *Journal of Australian Political Economy*, *Cultural Studies*, *Australian Feminist Studies* and *South Atlantic Quarterly*. She is currently working on a new book, *Home Economics: Speculating on Everyday Life*.

Nicole Cook is Lecturer in the School of Geography at the University of Melbourne, Australia. Nicole is an urban geographer with research interests in urban governance, power and participation, social movement and resident activism, housing and home. In 2011–2012 Nicole was project leader on the AHURI project 'The Impact of Third Party Objection and Appeal on Higher Density and Social Housing' (p30678) and in 2005–2006 she held a research fellowship on the project 'Banking on Housing: Spending the Home', supported by the ESRC/AHRC (RES-154-25-0012), University of Durham, UK.

Louise Crabtree is a Senior Research Fellow in the Institute for Culture and Society at Western Sydney University, Australia. Louise's research focuses on how property can be conceptualised and enacted to address social justice, and on participatory research methods. Both avenues are being used to simultaneously foster social innovation and equity outcomes on the ground, and explore and build theory on multi-stakeholder governance, decoloniality, property law, resilience and citizenship.

Aidan Davison is Senior Lecturer in human geography and environmental studies at the University of Tasmania, Australia. The author of 60 journal articles and book chapters, Aidan is fascinated and troubled by interdisciplinary questions of sustainability arising in the intersection of themes of nature, culture and technology, particularly in urban contexts.

His theoretical and qualitative research has covered topics such as suburban history, the urban forest, environmental movements and climate change.

Michelle Gabriel is a Senior Research Fellow in the Housing and Community Research Unit, University of Tasmania, Australia. She is currently undertaking research on sustainable housing, healthy ageing and housing, and monitoring reforms within Australia's social housing system. In relation to environmental sustainability, Michelle is evaluating a three-year trial of a community-based energy-efficiency program, Get Bill Smart. The project is in partnership with Sustainable Living Tasmania and Mission Australia and it is funded by the Australian Department of Industry. Michelle has also completed a national project, 'The Environmental Sustainability of Australia's Private Rental Housing Stock', funded by the Australian Housing and Urban Research Institute.

Andrew Gorman-Murray is a Senior Lecturer in Geography and Urban Studies at Western Sydney University, Australia. Housing, homemaking and domestic dynamics comprise one of his key research areas, particularly in regard to gender, sexuality, social identity and the domestic sphere. He has conducted ARC-funded work on LGBT experiences of home and on spatialities of masculinities and domesticities. He is also a practising artist, with a keen interest in exploring and developing visual languages of the domestic and other built environments through 'expanded photography'. His current work draws together cultural geography and contemporary art practice.

Cathy Keys is a Research Fellow in the Aboriginal Environments Research Centre at the University of Queensland, Brisbane, Australia. Her doctoral thesis, 'The architectural implications of Warlpiri jilimi', was concerned with the people–environment relations of Aboriginal women living in Central Australia. She is committed to exploring the social and cultural properties of architectural space.

Laurence Murphy is Professor of Human Geography at the University of Auckland, New Zealand where he was formerly a Professor of Property. An economic geographer by training, he has published widely on property topics including: home ownership, social rented housing, mortgage securitisation, office development, and entrepreneurial urban governance. He completed his PhD at the University of Dublin (Trinity College) and has held lecturing positions at the Queen's University Belfast and the London School of Economics. In 2014 he was appointed the Helen Cam Visiting Fellow at Girton College, Cambridge.

Jean Parker is a Researcher and Lecturer at Macquarie University and the University of Sydney, Australia. Part of her research has been published in the *Journal of Australian Political Economy*. Her doctorate explored Australia's policy response to the economic crisis of 2008 with a focus on neoliberalism and the welfare state. Her current research focuses on financialisation and housing policy.

Michael Rehm is a Senior Lecturer in Property at the University of Auckland Business School, New Zealand. His research focus has been on the development of hedonic pricing models used to isolate and better understand specific behaviour of market participants within housing and office markets. His research incorporates geographic information systems (GIS) to model and explore the often underappreciated spatial relationships prevalent in property markets, including work on school zoning, proximity to cell phone towers and leaky building stigma. His latest research is on housing affordability and the impacts of inclusionary zoning on residential land development.

Dallas Rogers is a member of the Institute for Culture and Society and a Lecturer within the School of Social Sciences and Psychology's Urban Research Program at Western Sydney University, Australia. Dallas' Institute for Culture and Society scholarship investigates the relationships between globalising urban space, discourse and technology networks, and poverty and wealth. He has undertaken a critical analysis of Australian urbanism through fine-grained empirical research with low-income urban citizenries, as well as super-rich transnational property investors and their agents. His current research interests focus on: a relational examination of housing poverty and wealth in globalising cities; foreign investment and the changing nature of Asian–Australian economic, technology and cultural relations; and the intersection between democracy, private sector development and state intervention.

Millie Rooney is a Research Fellow with the University of Tasmania, Australia. Her recent PhD examined sharing relationships between suburban Australian neighbours. Millie is an interdisciplinary scholar interested in Australian suburban lives, domestic practices and community connection. As well as having a scholarly interest in local community, Millie is an active member of her suburban sustainability group and local food cooperative.

Katrina Schlunke teaches and researches in cultural studies in the School of Communication at the University of Technology Sydney, Australia. Her research focus is on the intersections of Indigenous knowledges and materialisations of the past, queering the postcolonial and the more-than-human. She is co-editor of the *Cultural Studies Review* and her most recent publication was in *Courting Blakness* (Queensland University Press, 2015).

Wendy Steele is a Principal Research Fellow and Associate Professor in the Centre for Urban Research at RMIT, Melbourne, Australia. In 2012 she was awarded a three-year ARC DECRA fellowship and has recently published in journals such as *Housing, Theory and Society* and *Housing Studies*.

Elizabeth Jean Taylor is a McKenzie Post Doctoral Research Fellow in the Faculty of Architecture Building and Planning at the University of Melbourne, Australia. Her research interests include the spatial and socio-economic dynamics of housing markets and planning systems, particularly the use of third party objection and appeal rights and with a quantitative and spatial (GIS) focus. Elizabeth's post-doctoral work explores patterns of conflict around higher-density housing and other contentious land-use changes, including fast food, landfills and intensive farming. The extent of planning conflict over car parking is one theme of this work. She was awarded the 2012 Brian McLoughlin Award by the journal *Urban Policy and Research* for best ECR paper. Taylor has been a researcher on several AHURI-funded research projects dealing with the intersection of planning and housing supply, as well as a recent project for the Henry Halloran Trust exploring barriers to practitioner engagement with planning research.

Laurence Troy is a Research Associate at City Futures Research Centre at UNSW Australia and has a PhD in Urban Geography from the University of Sydney, Australia. His primary research interest is in the politics of urban sustainability and how cities can respond to environmental challenges of the twenty-first century. Laurence's PhD research focused on the political economies of urban change and its intersection with broader narratives on environment and sustainability policy in cities. Laurence's current research interests are in high-density housing, urban renewal and public transport.

Lorenzo Veracini is Associate Professor in History at Swinburne University of Technology, Australia. His most recent book, *The Settler Colonial Present*, was released in 2015. He is Editor-in-Chief of Settler Colonial Studies.

Phillipa Watson is a Research Fellow in the Housing and Community Research Unit, University of Tasmania, Australia. Phillipa has worked as a designer, a sustainability consultant and a researcher. In all roles she has sought to identify how people can create more sustainable built environments. In research, Phillipa investigates how householders make change towards more sustainable living conditions and practices. She is currently involved in a three-year project, 'Get Bill Smart', which explores the effects of support given to low-income householders to make energy-efficient changes for improved comfort, affordability and wellbeing.

1 The politics of housing/home

*Nicole Cook, Aidan Davison and
Louise Crabtree*

Opening inwards and outwards

> Coffee tables, moonbeams, credit and cash, hopes and fears, all
> interleave with the mobilisation and management of a core element of
> the materiality in owned housing.
>
> (Smith 2008, 11)

The diversity of elements in Susan Smith's conceptualisation of housing,
above, is striking. Rather than being a self-evidently discrete object, housing
composes a myriad of component parts; an assemblage of materials, affects,
ideas and finances that only come together through collaboration. While
drawing our focus to the quotidian, this foregrounding of the materiality
of housing is not a petition for the specific or the everyday. It is rather to
open up the house as a site that mediates between the particular and the
systemic. This always processual site is a meeting ground in which intensive
practices, materials and meanings tangle with extensive, financial,
environmental and political worlds. In these spaces the cultural activity
and meaning of being at home is inseparable from the techniques,
technologies and objects of housing. To encounter the house and home as
a site of interleaving and entangling, then, is to unbound it in two directions
at once: towards the concrete, the intimate and the experiential; and,
towards the general, the institutional and the collective. This unbounding
not only makes visible the continuities and inter-dependencies that exist
across the diverse range of professions and disciplines that participate in
the design, construction, investment, exchange, management and
representation of housing. It also highlights the irreducible capacity for
novel configurations of human dwelling.

   This collection embarks on the bi-fold motion of unbounding housing
and home. Specifically, it interrogates the coproduction of the materials,
meanings and practices of dwelling and worlds of finance, nature and power.
Our interest is to explore the making and unmaking of political, economic
and environmental relations in and through housing and home. Throughout
the text, contributors broadly focus on hybrid objects of inquiry, including

that of housing/home. We do so to explicitly challenge the traditional bounding of housing research with technical questions of matter, finance and policy (housing) and symbolic questions of meaning, identity and selfhood (home) (Jacobs and Malpas 2015; Jacobs and Smith 2008). That is, we seek to keep open the possibility of home as a physical and institutional manifestation, and the cultural and psychological possibilities of 'feeling at home' (Easthope 2004, 136) as a single question.

While housing research is richly diverse, a founding premise of this collection is that to address housing/home as a coproduction of diverse elements is to raise new normative, conceptual and empirical questions. Specifically, this approach extends the range of phenomena gathered in the achievement of housing/home, while also recognising the contribution of human dwelling to wider concerns of social and ecological wellbeing. In terms of expanding the scope of elements that contribute to housing/home, Susan Smith's work on owner-occupation in modern capitalist societies is indicative, showing that, as much as bricks and mortar, homes are constituted by flows of money, materials and affects (Smith 2008; Smith *et al.* 2006). This work has done much to open the 'financial black-box' of housing markets, increasingly understood as habit and feeling, as much as explicit calculation (see Smith *et al.* 2006). Inspired in part by science and technology studies, home has also been positioned as a socio-technical event: an uneven and fragile achievement held together through multiple policies, practices and materials (Jacobs 2006). A more-than-human lens has added animals, plants and microorganisms to the mix, marking a growing recognition of the role that non-humans play in the constitution and feeling of 'homeyness', and the limits of human agency that animal and plant architectures imply (Power 2005, 2009, 2012). Running alongside these more-than-human interventions, urban political ecologists (Heynen *et al.* 2006; Kaika 2004) and sustainability researchers in geography (Crabtree 2005, 2006a, 2006b; Davison 2006, 2011; Lovell and Smith 2010) have brought housing into contact with its technological and ecological outside, engaging with the resources and communities 'externalised' in the private home. Building on the conceptualisations of housing/home as an extensive network of materials and relations set in wider social and ecological worlds, the collection explores and troubles conventional distinctions between form and affect, markets and environments, materials and politics. For some chapters, this may mean working with the hybrid housing/home, while others unsettle and redraw the borders and boundaries of housing in other ways.

In unbounding the house/home, this book moves with the material and relational turns in social theory over the past three decades – associated with endeavours such as posthumanism, naturecultures, actor–network theory, assemblage theory, relational materialism and cosmopolitics (Bennett 2010; Braun and Whatmore 2010; Connolly 2013; Latour 2005). These innovations ground ontological questions of being and becoming in performative worlds of practice. In light of these conceptual openings, the contributions here variously consider housing/home as (often protracted and extensive) events within temporalities and spatialities established and maintained through webs of lived relation.

At home with power-relations

The work of unbounding housing/home is not, however, a retreat from concerns about uneven and unjust forms of economic and political power. Recent work in cultural geography, for example, has conceptualised a range of intimate domestic spaces and materials as part of broader political projects of identity, belonging and citizenship. Photographs, mementos and visual materials have been connected by Divya Tolia-Kelly (2004) to processes of decolonisation in the UK, affording British-Asian women a sense of home within the context of colonial histories. Similarly, home-repair and renovation are revealed as processes that for Allon (2008) underpin Australian nation-building and that for Gorman-Murray (2009, 2014) challenge the predominant political scripting of heteronormativity in Australia. Gender relations, too, are challenged in Cox's (2015) discussion of masculine home-making in New Zealand so that the space of renovation is a dynamic process of gender construction.

Building on these openings, the concern of this text is to explore the relationship between intimate practices and places of housing/home and the increasing entangling of housing with global capitalism, environmental change and colonial and neocolonial projects of dispossession. In capitalist, settler societies, such as Australia, interrogation of such relationships is surprisingly uncommon. Part of the reason for this neglect rests in the way that the normalising of owner-occupation, Australia's dominant tenure form, has enabled it to resist critical inquiry (see chapters by Crabtree; Davison). Public acceptance of owner-occupied housing can exclude and mask vital questions about the resources, economics, politics and social differences through which such housing is sustained. This includes: the resilience of market-models of housing against rising household debt; recognition of investors at the expense of residents on the margins of home-ownership; contemporary Indigenous dispossession and housing markets; and environmental sustainability and climate change. Recognising the significance of these interlinked questions, a core focus of this text is to open the myriad practices of housing/home to these wider political geographies.

The contention of this collection is that the intellectual, affective and bodily attachments that people have to particular manifestations of housing/home are integral to regimes of power. The aspirations of and risks faced by homeowners and potential homeowners are the forces that also marshal colonial, capitalist and other forms of power. Together these trajectories of inquiry provide a new lens through which the organising power of housing/home, and its tensions and contradictions, can be better apprehended. These strategies for unbounding housing/home penetrate more deeply into the norms and hegemonies in our time, providing insight into the ways these social realities take hold, and the ways they might then change. This mode of systemic inquiry is paradoxically both comprehensive and modest in its intent. By encompassing the actual complexity rolled and enrolled into housing/home, we are afforded a more realistic appraisal of possibilities for transformation, and more accurate account of what is at stake in political agendas for change.

Contributors to this volume employ a variety of strategies, from the purely theoretical to the highly empirical, and much in between, with which to unbound housing/home. To navigate the extensive, shifting and multiple networks that constitute and that are constituted by housing and home, while remaining true to the textures and contexts of actual houses and practices of home-making, we next draw out three key themes that run through this book. These themes are the identification of unchartered economies; environmental politics; and constitutive outsides in housing/home. Drawing on diverse disciplines (including from housing studies, geography, sociology, law and philosophy), we offer a partial synthesis of research under these themes and an account of the further inquiry this work provokes, before offering an overview of the contributions to this inquiry offered in this volume.

Unchartered economies of housing/home

One of the outcomes of encountering housing/home through socio-material and more-than-human geographies is a widening of the phenomena that might be recognised as participating in human dwelling. This material–relational turn unsettles the borders between the economic and the political, the social and the material and the cultural and the natural. For example, Smith *et al.* (2006), in their study of house prices in Edinburgh, found that house prices were derived not through cognitive and objective calculation, but through collectively produced forms of sensing or feeling. The attachments of owner-occupiers to goods and services purchased through *in situ* home equity withdrawal similarly reveal the affective and familial logics that drive mortgage-led consumption (Cook *et al.* 2013). While the risks of over-indebtedness should not be understated, this work found that 'debted objects' can hold homes and families together as much as pull them apart. Margaret Atwood (2008) argues more broadly that the ubiquity of housing debt in many modern societies exceed direct economic explanation, and is in part derived from the interplay of housing finance practices and narratives of risk and excitement.

Katherine Brickell (2012) traces the relationship between feelings of comfort and being at home with uneven economic development. In a literature that is usually preoccupied with the impact of geopolitics on the home, she calls for greater analysis of the ways that images and discourses of Western domesticity anchor uneven and unequal geopolitical relations. Harker's (2009) account of mortgage lending as a geopolitical regime is another example of the ways that everyday practices of mortgage consumption are entangled with the maintenance of political processes. To these insights, we add the ways that prosaic acts of home-making are also institutional, legal and financial acts that uphold different aspects of memory, identity and entitlement, and erase others. Crabtree (2013) develops this point in discussing opportunities for decolonising property in Australia, highlighting how particular forms of dwelling bound up in owner-occupation under capitalism often rest on and propagate a routine denial of Indigenous custodianship and occupation of land.

The interplay of human sensibilities and housing markets provides one of many examples of the presently unchartered economies of housing/home opened up by material–relational ontologies. This 'turn' throws light on inconsistencies and tensions produced in structures and markets conceived as habits as much as calculations, governed by affects of risk, security and hope, as much as supply and demand. It draws attention, too, to the striking – yet seemingly taken for granted – tensions in housing markets, such as the inelasticity of house prices and car parking (Taylor, this volume), and the uneven access to housing equity as welfare provision (Laurence and Rehm, this volume). The invitation here, then, is to unsettle conventional borders of housing and home so as to identify and interrogate the alliances across diverse fields that require investment and effort in order to sustain housing outcomes.

The environmental politics of housing/home

It is common to observe that housing and home are shaped by economic, environmental and political contexts, in the sense that these forces are taken to act on the domestic realm from the outside. Less common is to recognise the *production of* environmental logics through housing/home, that progress not through discrete sites but through dispersed systems that encompass everything from resource extraction, to outdoor recreation practices, environmental social movements, school curricula, urban population growth, wilderness photography, architectural fashions and building technologies.

For example, aesthetic sensibilities cultivated by concern about the growing scarcity of 'nature' in many modern societies have arguably gained expression in the price premiums and architectural logics now evident in much housing juxtaposed with urban nature, such as coasts, rivers, forests and mountains (Jim and Chen 2006; Randolph and Tice 2014). Housing thus enables limited and highly uneven social access to privileged (and predominantly scenic) encounters with urban natures.

From this perspective, our interest in this volume is to explore the simultaneous making of political, economic and environmental relations in and through housing and home.

More generally, housing/home is implicated in networks of resource use and waste production that bring with them a set of environmental politics. What are the forces that hold energy-, water- and material-intensive housing development in place? This implicates not just bodies and practices of homemaking and dwelling, but legal and property rights, water and energy infrastructure, ecological processes, natural landscapes, everyday practices and expert knowledges of maintenance and repair. Despite longstanding critique of the role of urban form, and housing type, on resource metabolism and waste production (Bunker and Holloway 2006; Newton and Meyer 2012; Quastel *et al.* 2012), and interest in homes as sites of new practices of sustainability (Lane and Gorman-Murray 2011), tensions between environmental outcomes and social outcomes associated with housing, dwelling and homemaking have barely been interrogated. Exceptions include studies of the complex policy dilemmas that result when environmental critiques of housing such as the 'McMansion' (Dowling and Power 2011) or low-density housing in traditional suburban neighbourhoods (Cook *et al.* 2013) are brought together with the social and personal values afforded by these housing forms, and Davison's (2008) study of the importance of suburban homeownership in the lives of environmentalists who are otherwise scathing about the impacts of suburban sprawl.

Explorations of the wider political and environmental webs of housing/home thus produce an uncomfortable politics, one that troubles the practices and materials of homemaking that contribute to wellbeing, security and comfort: to a sense of 'homeyness' (Power 2009). While the restorative aspects of home are by no means universal – they too are performed rather than given – the place of home, including the owned home, is more often than not tied to discourses of social wellbeing, stronger health outcomes and economic independence (Blunt and Dowling 2006). Practices of homemaking and investment are also often celebrated and divulged through a range of reality TV shows and consumer marketing strategies (see Allon 2008), while the business of being in debt can provide a narrative and focus for life (Atwood 2008).

The constitutive outside in housing/home

To recognise that housing/home requires the composition of diverse elements through collaborative processes is to confront contingency. Housing/home is made rather than given, performed rather than secured (Blomley 2013). In both housing and urban research, such collaborative emergence has been captured in the idea of assemblage (Cook *et al.* 2013; McFarlane 2011), which refers not only to the collaboration of unlike elements but the potential for such collaborations to widen capacities for change and innovation. Assemblages are also often characterised by their informality and improvisation, by their capacity to enable the 'unintended' use and reuse of spaces and materials in unexpected ways. McFarlane (2011) has used this framework to situate the informal settlement and slum housing at the centre of urban studies and housing research. While set apart from the attributes of formal cities – that is, without regulated forms of tenure, infrastructure or property rights – the slum nonetheless comes together to function as a city, often in unplanned ways and always against the odds. It does so by bringing all that is evacuated from the formal housing market – unpropertied land, consumer waste, 'left over' services, water, infrastructure – into structures and practices of dwelling.

To study housing/home as an assemblage is to interrogate the boundary between what lies inside and what lies outside the house and home. One way to ask how housing/home is constituted by its outside is to ask what housing and home would look like if all that is currently evacuated from traditional, formal, dominant modes of housing – all that currently constitutes its outside – were incorporated into the policies, practices and structures of dwelling (see Grosz 2001). Currently in Australia there is a resistance to forms of housing that do not conform to speculative logics of property investment under capitalism. The model of owner-occupation that results can exclude those on lowest incomes, younger people and Indigenous communities, among others. This exclusion is a central element of the economic and cultural processes of status and identity that constitute the meanings of homeownership. Outside this dominant housing model there exist many marginalised understandings of property, ownership and land (Crabtree 2008, 2013); and many lived experiences of dwelling. What, then, would housing and home look like if these excluded conceptions, practices and people (all of which are currently evacuated from formal housing) were incorporated into the institutional and structural logics of housing?

The open house/home in Australia

Exploring the three themes developed above, this text takes as its focus housing/home in Australia. With homeownership a cornerstone of the Australian settler and welfare state, this is an apposite and under-examined case through which to explore the complex politics of housing/home. While homeownership rates since the 1960s have been high, housing costs are an ongoing and intensifying issue. Debate continues as to whether the entry of first-home buyers into the market is being delayed or prevented (Bessant and Johnson 2013; Yates and Bradbury 2010); certainly, in 2015, investors briefly comprised a greater proportion of all new mortgages than owner-occupiers (Reserve Bank of Australia 2015). That debate, and concerns regarding the affordability and accessibility of ownership, run alongside persistent concerns regarding the expense and instability of Australia's private rental sector (Easthope 2014). Several factors underpin these issues, including the preferential tax treatment of owner-occupied and investment properties, alongside the ongoing residualisation of public rental housing (Phibbs and Young 2009). Consequently, there is growing interest in hybrid tenure options such as shared equity homeownership, cooperatives and community land trusts that unpack and rearrange the economic and social relationships of housing, to deliver stable and affordable housing options (e.g. Crabtree *et al.* 2015; Gilmour 2012).

Meanwhile, while market-rate ownership appears to slip beyond an increasing number of hands, discourses and policies of owner-occupation are increasingly promoted and utilised as a potential strategy for Indigenous self-determination (e.g., Council of Australian Governments Select Council on Housing and Homelessness 2013). Some Aboriginal and Torres Strait Islander individuals and communities are indeed seeking economic activity and gain through tenure reform: however, such households' and communities' core objectives in seeking 'ownership' vary immensely, and for some, issues of family and community autonomy and stability feature more prominently than expectations of wealth creation (Memmott *et al.* 2009). Rather than promoting a single tenure form, there is a challenge for discourse and policy focusing on Indigenous land tenure reform to manifest sufficient sensitivity and nuance to deal with the diversity of aspirations and tenure forms in these situations, including intersections with Native Title and community governance concerns (see Crabtree *et al.* 2015; Moran *et al.* 2010; Terrill 2016).

In addition, despite evident renewable energy resources, national policy reform in urban and environmental affairs has been slow in driving low-carbon and sustainable housing options, while residents themselves have pioneered new approaches towards sustainability (Crabtree and Hes 2009; Gabriel and Watson 2013; Strengers and Maller 2014). While owner-occupation has remained a central component of Australian urbanisation, welfare statism and citizenship, its central tenets and primary enactments are therefore challenged by the extant diversity of housing practices, aspirations and articulations.

Structure of the book

The chapters that follow explore different aspects of the unchartered economies, environmental politics and constitutive outside of housing and home. To these ends, the book is arranged into three parts. The first, 'Housing/home and worlds of finance', engages with housing markets as an interleaving of mortgage innovation, online technologies, private urban infrastructure (e.g., car parking) and planning. These diverse elements are assembled into particular constellations of social, political and economic possibility and constraint, while remaining largely invisible in dominant representations of housing markets.

The effects of online real estate technology on housing markets are considered by Dallas Rogers in 'Uploading real estate' (Chapter 2). These technologies move beyond online advertising of properties, to sites for real-estate professionals that link property sales data with institutional, taxation and immigration information across multiple nations. In addition to facilitating property-led immigration, such websites enable global speculation on housing, opening domestic markets more fully to global flows of finance. These technologies allow users to bid for a 'share' in a given housing market, while the management of the underlying asset is undertaken by a new stratum of real-estate investors. Through an analysis of Asian-Australian web-based innovation, and drawing upon philosophies of technology, Rogers positions the internet-based property website as an important actor in the development and evolution of contemporary housing markets, raising new questions about the intersection of digital technology, property investment and housing.

Socio-economic questions raised by the increasing fungibility of housing wealth are explored by Laurence Murphy and Michael Rehm in 'Homeownership, asset-based welfare and the actuarial subject' (Chapter 3). Unsettling the boundary between finance and materiality, this chapter is positioned in the context of the mortgage market deregulation that in many market societies enables homeowners to access housing equity without selling their home. Mortgage deregulation has played a key role in shifting the wider economic effects of owner-occupation and promoting debt-based welfare and consumption. Rather than a vehicle for the accumulation of housing wealth that can only be accessed by the sale of the physical structure, the owned home has become a resource for *in situ* consumption. Despite this dematerialising of housing wealth, housing equity is not evenly distributed by socio-economic status. The assemblage of home therefore raises significant policy issues in terms of social equity, particularly for those groups who may be expected to draw on this wealth to fund expenses in older age.

Chapter 4, 'Building on sand? Liquid housing wealth in an era of financialisation', by Fiona Allon and Jean Parker, also explores the institutional and policy settings that have facilitated the dematerialisation of housing. It is argued that these settings have simultaneously expanded the accumulation of risk by owner-occupiers as homes increasingly perform investment, insurance and spending roles. Recognising the access that owner-occupiers have to housing wealth *in situ*, the chapter documents how the unbounding of housing wealth from the physical structure of the home is the inevitable result of public policies that fuel property speculation. These policies reflect a longer-term shift from welfare to 'wealthfare' states, in which housing wealth is employed as a solution to the crisis of the welfare state.

Despite the evident flexibility of housing markets, Chapter 5, 'Cohabiting with cars', by Elizabeth Taylor, reveals their paradoxical inertia in relation to Australian car dependency. Through an analysis of housing and transport data, planning controls and historical trends in urbanisation, Taylor exposes the myriad of institutional, legal and material norms that bind owned homes to car parking spaces. Even in the face of new demand for housing without car spaces, the links between housing and car parking remain tenacious. While revealing the hidden role of car parking in shaping Australian housing markets and urbanisation, the chapter also recognises a plurality of housing–transport relations that might be cultivated. To this end the chapter presents a case study of the controversial car-free Nightingale housing development in Melbourne.

Part II, 'Housing/home and worlds of nature', presents work exploring the more-than-human relations of housing/home and the related political

10   *Nicole Cook* et al.

deployments of nature and sustainability discourses within cities. In Chapter 6, 'Secure in the privacy of your own nature', Aidan Davison builds on research in urban political ecology to investigate the relationship between urban nature and suburban owner-occupation in Australian cities. This chapter explores the paradoxical coproduction of suburban-led capital accumulation, consumption and pollution with private yearning for restorative contact with nature in the forms of private gardens, urban forests, suburban coastlines and mountains and ready escape into nonurban landscapes. More-than-human realities are enrolled in the creation of heavily fortified boundaries around the private home that have enabled Australians to participate in a capitalist-colonial project of modernisation while seeking to live at a safe distance from the many negative consequences of this project.

Laurence Troy's 'Making nature and money', Chapter 7, also engages with urban political ecology to consider the relationship of housing and urban nature, although in this case the focus is on the urban development and planning process. The chapter is situated at the intersection between discourses of urban renewal, waterfront redevelopment and nature in East Perth. While natural features such as waterways and public parks feature in the redevelopment of such sites, Troy argues that the potential for socially and ecologically sustainable outcomes are limited by three powerful discourses of nature advanced by the development authority: that the site was empty and waiting for renewal; that such renewal is best undertaken to attract urban elites; and that the development of public spaces be undertaken with a goal of maximising capital accumulation through private property markets. Despite the myriad of naturecultures that such development might advance, these are ultimately limited by the hegemonic subordination of nature to capital that underpins the planning and development process.

Chapter 8, 'Displacement as method', by Lorenzo Veracini, offers a study of three disparate social movements that each articulate radical political dissent through housing/home. Motivated variously by a concoction of libertarian and environmental values, the seasteading, tiny houses and 'Freeman on the Land' movements share a characteristic settler colonial method of expressing this dissent that seeks to avoid formal political institutions through strategies of spatial displacement. Veracini uncovers the way in which these movements attempt to avoid or modify the regulatory and financial controls around owner-occupied housing, unsettling boundaries between residents, property and housing. Calling into question the relationship between domestic sovereignty, as expressed through capitalist land and house ownership in Australia and other settler colonial societies, and the political sovereignty of the nation-state, these movements expose the role of housing/home in the power-laden and more-than-human inscription of territorial boundaries.

In Chapter 9, '"The Best House Possible"', Michelle Gabriel, Millie Rooney and Phillipa Watson displace familiar instrumentalist accounts of housing energy efficiency with a textured qualitative account of the practices of 'home comfort'. This unsettles the boundary between biophysical accounts of thermal comfort and cultural narratives of the comforts of home. This unsettling is deepened by the juxtaposition of home comfort with the uncomfortable manifestation of social disadvantage in housing. Developing a case study of Get Bill Smart, a government energy-efficiency program for low-income households in Hobart, Tasmania, the chapter approaches housing/home through social practice theory. This recognises that home comfort is an ongoing relational achievement drawing together extensive technological systems, the material fabric of the house, the inhabitants of home, and the activities of living into a dynamic assemblage. Exploring the interactions of human inhabitants with each other and with nonhumans, technologies and institutions, the authors conclude that residents' experience was poorly understood by government agencies focused on linking agendas of social welfare and environmental sustainability.

Part III, 'Housing/home and worlds of possibility', explores the potential for innovation and transformation of housing and home focused on wider political, economic and environmental processes. These chapters explore the productive capacities of citizens, materials and spaces to reconfigure the boundaries of housing/home in Australia in ways that support human and more-than-human flourishing. In Chapter 10, 'Unbounding home ownership in Australia', Louise Crabtree expands the concept of 'citizen virtue' beyond the limited tenure of owner-occupation that has dominated Australian cultural and economic life, instead recognising the diverse contributions and attachments to place fostered by those outside mainstream housing markets. Drawing on the diverse relationships to place cultivated through social housing tenancy in Millers Point in Sydney and on Aboriginal and Torres Strait Islander lands, this chapter shows that dominant narratives of owner-occupation as a preferred or desirable housing system quickly unravel. While such narratives can be seen as mechanisms to police acceptable housing behaviours and thus citizenship, the persistence of non-conforming tenures present a disjuncture in housing policy that may offer a productive space for the dislocation of citizen virtue from the owner-occupied home.

In Chapter 11, 'Performing housing affordability', Nicole Cook positions urban protest in cities as potential sites of innovation and experimentation in housing policy. Through a re-reading of Sydney's Green Bans (1971–1974), this chapter argues that urban social movements can play important roles in urban and housing policy outcomes. Providing a companion piece to the contemporary tensions in Millers Point discussed in Chapter 10, this chapter explores this site in the 1970s, when it was one of around 40 urban development sites subject to Green Bans. While long recognised as a cross-class, grassroots movement in opposition to urban development, the Green Bans are presented here as an important force in the modification of urban policy to incorporate affordable housing in waterfront redevelopment processes. Central to the effectiveness of this activism was the capacity to interrupt the urban development process that turned building sites into incubators of socially and ecologically sustainable development processes. While such occupation is difficult to achieve, its effectiveness provides unique insights into the 'vital materialisms' that contribute to both the enduring power and the efforts to control the occupation of cities by citizens and workers.

In Chapter 12, 'Interstitial housing space: no centre just borders', Wendy Steele and Cathy Keys begin with housing materials and spaces that evade singular value or meaning. They focus in particular on the domestic spaces of 'under the house' in the Queensland colonial vernacular, and the yulka of remote Aboriginal (Walpiri) camps in Central Australia. These are spaces whose uses, meanings and values are open to experimentation and innovation. Used for a myriad of purposes, these spaces are likely to be lost through processes of land intensification or overlooked in favour of the privatisation of housing. Yet strategies of investment and privatisation are ill-equipped for the ebb and flow of daily life: the Queenslander is prone to flooding, and the Walpiri camps are communal rather than private dwelling spaces. In such contexts the indeterminacy of these interstitial spaces provides important resources for negotiating social and ecological uncertainty.

Katrina Schlunke's 'Burnt houses and the haunted home' completes the collection in Chapter 13 by unsettling the boundary between the material house and the affective home through the figure of the ruin. Drawing upon first-hand experience of 'losing a house' in a bushfire, the author offers a critique of the Australian reflex of erasing ruined houses as rapidly as possible in post-disaster landscapes. Rather than presenting the ruin as the negation of housing/home, Schlunke explores the capacity of the ruin to heighten sensitivity to the relational web of matter, memory and meaning in the experience of home. She explores the capacity of the ruin to reconfigure settler Australians within colonial legacies of dispossession and denial. The ruin is presented as a site of temporal and spatial possibility in which can arise new ethical relations to place and possession.

Each of the three parts of the book begin with a selection of four images from Andrew Gorman-Murray's photographic artwork, Thrown-Togetherness, 2015. The intellectual and aesthetic motivation underlying this creative response to the call for papers on which this volume is based is discussed in a short exegesis at the end of the book. Gorman-Murray, a cultural geographer, offers a critically engaged visual exploration of the heterogeneous and performative constitution of external, internal and interstitial spaces in housing/home. Hence the work comprises three series: the External, the Internal, and the Interstitial, and images from each series have been selected for inclusion to evoke and expand upon the themes of each part, and of the overall collection.

References

Allon, F. 2008. *Renovation Nation: Our Obsession with Home*. University of New South Wales Press, Melbourne.

Atwood, M. 2008. *Payback: Debt and the Shadow Side of Wealth*. Harper Collins, Toronto.

Bennett, J. 2010. *Vibrant Matter: A Political Ecology of Things*. Duke University Press, Durham.

Bessant, J.C., and Johnson, G. 2013. 'Dream on': declining homeownership among young people in Australia? *Housing, Theory and Society* 30(2): 177–192.

Blomley, N. 2013. Performing property: making the world. *Canadian Journal of Law and Jurisprudence* 26(1): 23–48.

Blunt, A. and Dowling, R. 2006. *Home*. Routledge, London.

Braun, B. and Whatmore, S. (eds) 2010. *Political Life: Technoscience, Democracy and Public Life*. University of Minnesota Press, Minneapolis.

Brickell, K. 2012. Geopolitics of home. *Geography Compass* 6(10): 575–588.

Bunker, R. and Holloway, D. 2006. Planning, housing and energy use: a review. *Urban Policy and Research* 24(1): 115–126.

Connolly, W.E. 2013. *The Fragility of Things: Self-organizing Processes, Neoliberal Fantasies and Democratic Activism*. Duke University Press, Durham.

Cook, N., Smith, S.J. and Searle, B.A. 2013. Debted objects: homemaking in an era of mortgage-enabled consumption. *Housing Theory and Society* 30(3): 1–19.

Cook, N., Taylor, E. and Hurley, J. 2013. At home with strategic planning: reconciling resident attachments to home with policies of residential densification. *Australian Planner* 50(2): 130–137.

Council of Australian Governments Select Council on Housing and Homelessness. 2013. *Indigenous Home Ownership Paper*. Council of Australian Governments, Canberra.

Cox, R. 2015. Materials, skills and gender identities: men, women and home improvement practices in New Zealand. *Gender, Place and Culture*. DOI:10.108 0/0966369X.2015.1034248

Crabtree, L. 2005. Sustainable housing development in urban Australia: exploring obstacles to and opportunities for ecocity efforts. *Australian Geographer* 36(3): 333–350.

Crabtree, L. 2006a. Sustainability begins at home? An ecological exploration of sub/urban Australian community-focused housing initiatives. *Geoforum* 37(4): 519–535.

Crabtree, L. 2006b. Disintegrated houses: exploring ecofeminist housing and urban design options. *Antipode* 38(4): 711–734.

Crabtree, L. 2008. The role of tenure, work and cooperativism in sustainable urban livelihoods. *Acme* 7(2): 260–282.

Crabtree, L. 2013. Decolonising property: exploring ethics, land, and time, through housing interventions in contemporary Australia. *Environment and Planning D: Society and Space* 31(1): 99–115.

Crabtree, L. and Hes, D. 2009. Sustainability uptake in housing in metropolitan Australia: an institutional problem, not a technological one. *Housing Studies* 24(2): 203–224.

Crabtree, L., Moore, N., Phibbs, P., Blunden, H. and Sappideen, C. 2015. *Community Land Trusts and Indigenous Communities: From Strategies to Outcomes*. Final Report No. 239. Australian Housing and Urban Research Institute, Melbourne.

Davison, A. 2006. Stuck in a cul-de-sac? Suburban history and urban sustainability in Australia. *Urban Policy and Research* 24(2): 201–216.

Davison, A. 2008. The trouble with nature: ambivalence in the lives of urban Australian environmentalists. *Geoforum* 39: 1284–1295.

Davison, A. 2011. A domestic twist on the eco-efficiency turn: technology, environmentalism, home. In R. Lane and A. Gorman-Murray (eds) *Material Geographies of Household Sustainability*. Ashgate, London.

Easthope, H. 2004. A place called home. *Housing Theory and Society* 21(3): 128–138.

Easthope, H. 2014. Making a rental property home. *Housing Studies* 29(5): 579–596.

Gabriel, M. and Watson, P. 2013. From modern housing to sustainable suburbia: how occupants and their dwellings are adapting to reduce home energy consumption. *Housing, Theory and Society* 30(3): 219–236.

Gilmour, T. 2012. *We're all Landlords and Tenants*. Retrieved 4 January 2016 from http://tonygilmour.com/yahoo_site_admin/assets/docs/CENSW_report.308232112.pdf

Grosz, E. 2001. *Architecture from the Outside: Essays on Virtual and Real Space*. MIT Press, Boston.

Harker, C. 2009. Spacing Palestine through the home. *Transactions of the Institute of British Geographers* 34(3): 320–332.

Heynen, N., Kaika, M. and Swyngedouw, E. (eds) 2006a. *In the Nature of Cities: Urban Political Ecology and the Politics of Urban Metabolism*. Routledge, London and New York.

Jacobs, J.M. 2006. A geography of big things. *Cultural Geographies* 13(1): 1–27.

Jacobs, K. and Malpas, J. 2015. Material objects, identity and the home: towards a relational housing research agenda. *Housing, Theory and Society* 30(3): 281–292.

Jacobs, J.M. and Smith, S.J. 2008. Living room: rematerialising home. *Environment and Planning A* 40: 515–519

Jim, C.Y. and Chen, W.Y. 2006. Impacts of urban environmental elements on residential housing prices in Guangzhou (China). *Landscape and Urban Planning* 78(4): 422–434.

Kaika, M. 2004. Interrogating the geographies of the familiar: domesticating nature and constructing the autonomy of the modern home. *International Journal of Urban and Regional Research* 28(2): 265–286.

Lane, R. and Gorman-Murray, A. (eds) 2011. *Material Geographies of Household Sustainability*. Ashgate, London.

Latour, B. 2005. *Reassembling the Social: An Introduction to Actor–Network Theory*. Oxford University Press, Oxford.

Lovell, H. and Smith S.J. 2010. Agencement in housing markets: the case of the UK construction industry. *Geoforum* 41: 457–468.

McFarlane, C. 2011. The city as assemblage: dwelling and urban space. *Environment and Planning D Society and Space* 29(4): 649–671.

Memmott, P., Moran, M., Birdsall-Jones, C., Fantin, S., Kreutz, A., Godwin, J., Burgess, A., Thomson, L. and Sheppard, L. 2009. *Indigenous Home-Ownership on Communal Title Lands*. AHURI Final Report No. 139. Australian Housing and Centre, Melbourne.

Moran, M., McQueen, K. and Szava, A. 2010. Perceptions of home ownership among Indigenous home owners. *Urban Policy & Research* 28(3): 311–325.

Newton, P. and Meyer, D., 2012. The determinants of urban resource consumption. *Environment and Behaviour* 44(1): 107–135.

Phibbs, P. and Young, P. 2009. Going once, going twice: a short history of public housing in Australia. In S. Glynn (ed.) *Where the Other Half Lives: Lower Income Housing in a Neoliberal World*. Pluto Press, London.

Power, E. 2005. Human–nature relations in suburban gardens. *Australian Geographer* 36(1): 39–53.

Power, E. 2009. Border-processes and homemaking: encounters with possums in suburban Australian homes. *Cultural Geographies* 16(1): 29–54.

Power, E. 2012. Domestication and the dog: embodying home. *Area* 44(3): 371–378.

Quastel, N., Moos, M. and Lynch, N. 2012. Sustainability-as-density and the return of the social: the case of Vancouver, British Colombia. *Urban Geography* 33(7): 1055–1084.

Randolph, B. and Tice, A. 2014 Suburbanizing disadvantage in Australian cities: sociospatial change in an era of neoliberalism. *Journal of Urban Affairs* 36(S1): 386–399.

Reserve Bank of Australia 2015. *Submission to the Inquiry into Home Ownership – June 2015*. Retrieved 4 January 2016 from www.rba.gov.au/publications/submissions/ inquiry-into-home-ownership/proportion-investment-housing-relative-owner-occ-housing.html.

Seamon, D. 1993. *Dwelling, Seeing, and Designing: Toward a Phenomenological Ecology*. State University of New York Press, Albany, NY.

Smith, S.J. 2008. Owner-occupation: at home with a hybrid of money and materials. *Environment and Planning A* 40(3): 520–535.

Smith, S.J. and Munro, M. 2009. *The Microstructures of Housing Markets*. Routledge, Abingdon.

Smith, S.J., Munro, M. and Christie, H. 2006. Performing (housing) markets. *Urban Studies* 43(2): 83–98.

Strengers, Y. and Maller, C. (eds) 2014. *Social Practices, Intervention and Sustainability: Beyond Behaviour Change*. Routledge, London.

Terrill, L. 2016. *Beyond Communal and Individual Ownership: Indigenous Land Reform in Australia*. Routledge, London.

Tolia-Kelly, D.P. 2004. Materializing post-colonial geographies: examining the textural landscapes of migration in the South Asian home. *Geoforum* 34: 675–688.

Yates, J. and Bradbury, B. 2010. Home ownership as a (crumbling) fourth pillar of social insurance in Australia. *Journal of Housing and the Built Environment* 25(2): 193–211.

Part I

Housing/home and worlds of finance

Andrew Gorman-Murray, *The External 4*, 2015.

Andrew Gorman-Murray, *The External 6*, 2015.

Andrew Gorman-Murray, *The External 7*, 2015.

Andrew Gorman-Murray, *The Interstitial 1*, 2015.

2 Uploading real estate
Home as a digital, global commodity

Dallas Rogers

Introduction

Susan Smith (2008, 521) conceptualizes the home as a hybrid of mobile capital, physical materials and subjective lived experiences. Digital real estate platforms are increasingly playing a key role in these relationships; they not only shape but can also rupture the entanglements between the flow of cash, the materiality of real estate and the more subjective use values of home. A good example of the way that digital real estate platforms are reshaping home is BrickX, a small-scale real estate technology (tech) start-up in Sydney, Australia. Their promotional video, quoted verbatim below, strips the idea of fractional real estate investment from its complex financial industry moorings, leaving it in its most elementary form.

> BrickX: The new way to enter the property market.
> BrickX buys properties in prime locations.
> Then divides each property into 10,000 'Bricks'.
> A 'Brick' = A fraction of a property.
> A $1 Million property = $100 per 'Brick'.
> 'Bricks' start at $66.
> Invest in 'Bricks' across a range of properties.
> Earn monthly distributions from rental income.
> Share the capital returns.
> While we take care of the property management.
> BrickX: The new way to enter the property market.
> What are you waiting for?
>
> (BrickX, 2015, how-it-works)

Complex fractional financial practices underwrote the subprime mortgage implosion that resulted in a global financial crisis (Dufty-Jones 2016). The more primitive fractional 'financial alchemy' (Ferguson 2008, 270) of BrickX breaks the dwelling down into a set of constituent parts that can be traded independently of the home itself. The use of fractional financial alchemy, argues Ferguson (2008, 254), 'is not so much about real estate as about surreal estate' whereby the constituent parts of a dwelling are detached from the subjective use values of the home. These constituent parts are commodified and sold to the highest bidder; there could be thousands of kilometres between those living in the dwelling, the mortgage borrower (e.g. BrickX) and the fractional surreal estate investors.

BrickX (2015) identifies strongly with the real estate tech industry, and particularly with the industry discourse of 'disruptive technology'. It is surprising, however, that the digital uploading of residential real estate onto internet-enabled real estate technologies has not attracted more attention. We are at the beginning of a digitally driven, global expansion of the residential real estate industry. Millions of local residential homes from at least a quarter of the countries around the world have been uploaded onto the internet for sale as global commodities to be traded by a new stratum of global real estate professionals and investors (Juwai 2014; Rogers 2016a; StreetSine Technology Group 2014). The globalising real estate industry is increasingly comprised of national and international real estate sales agents, property developers, financial advisers, home loan brokers, foreign investment lawyers, immigration consultants and information technology professionals (Rogers 2016a; Rogers *et al.* 2015). While collecting empirical data for a study on Chinese investment in Australian real estate, I encountered a set of conceptual problems relating to the digitisation of the real estate industry across the Asia-Pacific. Some tech professionals were claiming that the unboundedness of their 'innovative technology solutions to property investing' would 'disrupt the [real estate] status quo' (BrickX 2015, about). The digital functionality and global scale of this new real estate industry forces us – conceptually, methodologically and empirically – to move beyond the nation-state centrism that frames much of the debate about domestic and foreign real estate investors (Tiwari and White 2010).

The boundaries of real estate and nation-state, money and materials, people and homes, are being rearticulated through internet-enabled real estate technologies (Rogers 2016a). Real estate technologies are central to the operation and interconnection of global real estate professionals and businesses across different legal, spatial, cultural, linguistic and technological frontiers (Isin and Ruppert 2015; Rogers *et al.* 2015). To take a regional example, the Australian-backed but Asian-based real estate tech start-up Juwai (2014) is one of the largest international real estate websites operating in the Asia-Pacific. The Australian co-founder and co-CEO stated that he 'realised that large numbers of [mainland] Chinese were buying property in Hong Kong [and he asked himself] … so where will these buyers go next?' He 'realized the potential for an international property portal for Chinese buyers. After a year of … doing research and focus groups in China and building our website, we launched in 2011' (Millward 2014, 1). The Mandarin word Juwai translates as 'home overseas' and the company's core business is to advertise foreign real estate and to procure real estate sales across nation-state boundaries for real estate companies from around the world.

The emergence of global real estate tech companies, like Juwai, opens up a suite of new housing research questions. For example, what will be the effect, if any, on the users of Juwai's global real estate technology when they access the 'most integrated platform connecting international agents and Chinese buyers' on their tablet or smartphone? Juwai (2014, 1) state that their tech product has '2.4 million property listings from 58 countries, giving Chinese buyers the most comprehensive collection of overseas property to search from'. What will be the long-term domestic and foreign housing market effects, if any, if the next generation of Chinese and other nationals power-up a real estate tech product on their smartphone to look for international real estate? I do not propose to answer these types of questions here. Rather, I pose these questions to clear a conceptual space to begin to think about how the globalising real estate tech products – which will increasingly frame how people from many countries around the world will buy and sell residential real estate – might come to confirm, shape or remould different understandings about land, real estate, home, citizenship and property. Indeed, I set out here to ask a more fundamental question about the unbounding of real estate and home. What are the political implications of the digital commodification of real estate, which involves detaching the material dwelling from the subjective lived experience of habitation, into digital data to facilitate new types of capital circulation and accumulation?

Analysing the uploading of real estate

The sections that follow think through these questions concerning digital real estate technologies by focusing on the 'technics' (rules of knowledge production) of real estate (Ortega 1939). First, a brief history of the internet highlights the ideological landscape that frames the real estate tech industry. This discussion accounts for small real estate data, which refers to information that, with limited analysis or technological manipulation, is ready for human comprehension (Manyika *et al.* 2011, 4). It also accounts for big real estate data sets, or what Burrow and Savage (2014) have called 'large scale digital data'. Manyika *et al.* (2011, 4) argue that 'big data has now reached every sector in the global economy', and this is increasingly the case with the globalising real estate industry. Isin and Ruppert (2015, 26) argue that new socio-technical relationships between people, technological objects and data enable a special class of digital 'speech act' that cuts across these data types, and what is at stake in these digital acts is 'the production, dissemination, and legitimation of knowledge'. Therefore, second, and more substantively, the analysis turns to the real estate technics and ideologies that are being uploaded into the digital and globalising real estate network. It is clear that libertarian, private property and market-based ideologies are central to real estate tech entrepreneurship. When exploring the ideological enframing of these digital technics it is hard to avoid the allure of Jacques Ellul's (1954) seminal neo-Marxist historical materialist study of technology. Ellul (1981,

155) wrote reflectively in 1981, 'if Marx were alive in 1940 he would no longer study economics or the capitalist structures but technology'. I suspect that a tech-savvy Marx would have taken an interest in the BrickX fractional real estate investment platform, and particularly the digital alienation of the investors from the use value of the properties they are investing in. It is not so much the real estate technology that is important here, but rather the real estate technics that are uploaded into the technology. As I show below, the entrepreneurs and tech companies are building particular ideologies into the materiality (i.e., software and hardware) of their real estate platforms. These material manifestations of ideology are designed to flow across the different national and international legal frameworks that frame transnational real estate sales. They are also built to flow across the cultural and linguistic frontiers to enable transcultural real estate sales.

Ellul's corpus is useful for exposing how free-market, capitalist and libertarian ideologies are deeply buried within the tech industry and their real estate tech products, even into the notion of real estate itself. The limitation with Ellul's work is that it is profoundly determinist. Technological determinism, to paraphrase Friedrich Kittler's (2002) famous claim, is the belief that technological media determine our human condition. While Kittler captures, at least intuitively, a part of the contemporary experience of modern communications technologies, determinist thinking, in all its forms, has been heavily critiqued for rupturing the agency of individuals and other social actors by pre-determining human pasts, presents and futures. Marxism, and particularly Marx's form of historical materialism, has also been accused of this type of teleology. Winthrop-Young (2011) argues that Kittler's work is more indebted to the discursive scholarship of Michel Foucault than the determinist historical materialism of Karl Marx. Winthrop-Young (2011, 145) argues that Kittler's statement, that technological 'media determine our situation', is a highly qualified statement that does not presuppose a *telos*. Kittler (2002, 153) argues that 'technological media are never the inventions of individual geniuses, but rather they are a chain of assemblages that are sometimes shot down and that sometimes crystallise'. José Ortega y Gasset's (1939) seminal work on the concept of technics offers a concise survey of the historicity of thinking through the technological techniques or arts of technology. Initially, a technic is discovered, and as Kittler suggests, this often involves many actors and occurs through trial and error, or even by accident. Through repeated use, a technic becomes a conscious practice, and if successful it might be handed down from one generation to another. Thus, the third step in the argument is to deploy Ortega and Kittler's conceptualisation of technics and technological media to undertake a historical analysis of the *technics of the real estate technicians* and the *technics of the real estate industries*. This analysis exposes some surprising ruptures and continuities. One of the most compelling continuities is that the real estate tech entrepreneurs are uploading the real estate technics that created significant housing inequity in the twentieth century into their twenty-first-century platforms. A key rupture is the emergence of big real estate data, which will increasingly be a sought after and perhaps even tradable commodity.

Isin and Ruppert's (2015, 34) work usefully prompts questions such as: who owns, has access to and is restricted from using the growing volume of real estate data? Who profits from these data, and will some large real estate tech companies seek to control the global real estate data market? Technology theorists have long pondered the dangers of an unfettered commitment to the modern technological project, and they offer different tools for theorising and exposing the embedding of free-market, capitalist and libertarian ideologies within real estate tech products. Martin Heidegger (1977), who is also 'not a determinist' (Lovitt 2013, xiii), famously rejects an instrumental view of technology, such as engineering, that attempts to position technology as an applied science. Heidegger argues that modern technologies are a 'revealing' or 'enframing' (*Ge-stell*), and an exploitation of nature and natural resources (*Bestand*) in the world. For Heidegger, modern technologies create a world of resources to be used and consumed, and this technic is premised on the objectification and quantification of the natural world, which leaves out the 'thinghood' in things. This creates the conditions for producing objects, including landed property and real estate. Jean-Jacques Rousseau (1750) calls into question the Enlightenment idea that technological and scientific progress will necessarily lead to the advancement of society by unifying technological knowledge and wealth with a virtue ethics. For Rousseau and Heidegger, humanity, or what it means to be human, and our subjective realities are intricately bound up with technics and technological objects. In their respective seminal texts, Lewis Mumford (1934) and Ellul (1954) also raise ethical and moral questions to warn that technologies could be used to exercise punitive forms of power over others, or to narrow human experience. Mumford (1934), in particular, argues that this should be challenged and guarded against. Therefore, the chapter concludes by offering some further sites for critical housing scholarship in relation to the global digitisation of the residential real estate industry.

Libertarianism and the real estate tech industry

Considering first the ideological landscape that frames the tech industry, Isin and Ruppert (2015, 34) argue that 'power relations in contemporary societies are being increasingly mediated and constituted through computer networks that eventually came to be known as the Internet'. There is a weighty theoretical corpus devoted to people, technology and societal relationality (Burrows and Savage 2014; Ellul 1954; Mitcham 1994; Tufeki 2014). However, the history of the world wide web and the internet (Keen 2015), and the developing real estate tech industry (Rogers 2016a), are oft-ignored contextual frames for thinking through twenty-first-century real estate relations.

In 1996, seven years after Tim Berners-Lee invented the world wide web, Robert Franks and Philip Cook (1996) argued that information technologies were poised to exacerbate global economic inequality. In the same year, John Perry Barlow (1996) published *A Declaration of the Independence of Cyberspace* to discursively frame the rise of the internet. The opening of Barlow's (1996, 1) declaration was ideologically instructive:

> Governments of the Industrial World … I come from Cyberspace, the new home of Mind…. You are not welcome among us…. Governments derive their just powers from the consent of the governed. You have neither solicited nor received ours…. Cyberspace does not lie within your borders. Do not think that you can build it, as though it were a public construction project. You cannot…. You have not engaged in our great and gathering conversation, nor did you create the wealth of our marketplaces. You do not know our culture, our ethics, or the unwritten codes that already provide our society more order than could be obtained by any of your impositions.

Barlow's libertarian proclamation provides revealing insights into the early thinking of information technologists. At the end of the twentieth century, these technologists were interested in moving from the 'industrial' into the 'information' age, and thereby denounced geographically bounded notions of governance and citizenship (Isin and Ruppert 2015). They proclaimed the independence of cyberspace in an attempt to reconstitute it as an emerging marketplace *space* that could be decoupled from government regulation and taxation. Cyberspace, occurring to Barlow and his contemporary tech followers, should be self-regulated through a set of independent tech industry cultural and ethical norms. This type of libertarian enframing of technology was not unexpected. Carl Mitcham (1994) was writing about it in his *magnum opus* on technology in 1994. He argued in 'the background of virtually all science and technology studies there lurks an uneasiness regarding the popular belief in the unqualified moral probity and clarity of the modern technological project' (Mitcham 1994, 1). Twenty years later, Andrew Keen (2015), one of Silicon Valley's most vocal contemporary critical insiders, is still uneasy. Keen argues (2015, 189) that a 'disruptive libertarianism', which he describes as a free-market liberalism that is bolstered by a 'disdain for hierarchy and authority, especially the traditional role of government', persists as one of the central ideological reference points in the tech industry. This free-market libertarianist thinking is the guiding ideology at tech conferences and large tech companies in the United States, and has been built into the paywalls of the tech industry's products (Isin and Ruppert 2015; Keen 2015).

Keen (2015, 228) redeploys Silicon Valley's own lexicon back onto itself to conclude that the internet-enabled tech industry is an 'epic fucking failure'. In the tech and venture capital industries the failure of your first tech start-up is celebrated as an entrepreneurial milestone.

The tech industry is not an entrepreneurial failure, but rather a democratic failure. Keen's history of the internet demonstrates that the utopian dreams for an open access and decentralised, but state funded, world wide web of the mid-twentieth century were subsumed into the hegemonic, hierarchical and monopolistic private tech sector of the twenty-first century. Isin and Ruppert (2015, 28) describe cyberspace as 'a space of relations between and among bodies acting through the Internet'; the early phase of cyberspace development was 'primarily a story of how the Internet was invented for national security and civic goals. It's a story about how public money … paid to build a global electronic network', argues Keen (2015, 38). The economist Mariana Mazzucato (2013) points out that the most successful tech companies received public sector start-up funding (e.g. Apple) or university scholarship funding for the founding CEOs (e.g. Google). Before 1991, the US government maintained legal control of the world wide web and companies that sought access to it were required to limit their use to 'research and education', albeit, as Friedrich Kittler (1995) has shown, this often meant military research and development.

Reflecting on the rise of the world wide web and internet, and the free-market liberalism and disdain for government that information technologists deployed to frame its emergence, it is perhaps not surprising that real estate was so readily uploaded into this technological system. According to some real estate industry sources (Movoto 2014), real estate listings periodically appeared on the internet from about 1994. At the turn of the century, and only a decade after the creation of the world wide web, one of the first Web 1.0 real estate companies 'was founded on the belief that selling or buying a home could be faster, easier and more efficient' (Zipreality 2014). In the same year Move (formally HomeStore) became one of the first internet real estate companies to be publicly listed. Web 1.0 was one of the initial large-scale software paradigms on the internet. This platform is typified by businesses that upload their core business functions onto the internet in a fairly straightforward manner. The assumption behind Web 1.0 technologies is that a company's non-virtual technics will operate in the virtual spaces of the internet in much the same way as they operated before the internet. Uploading the practice of mailing letters onto an internet platform as email, through a service provider such as Yahoo, is a good example of this type of Web 1.0 uploading practice.

Throughout the first decade of the twenty-first century the Web 1.0 uploading of real estate practices, or taking the old real estate technics online, increased significantly. A tech developer from a real estate tech start-up I interviewed in Asia in 2014 stated:

> Everything is going mobile of course, and we've benefitted tremendously … within our first year the smartphone came out. And then we jumped on that. And so we really became an app builder concentrating on property, so managing the information that's there.
>
> (Rogers 2016a, 7)

Large sections of the real estate industry are still operating in the Web 1.0 space. Real estate companies are contracting tech companies to help them upload their core business technics onto the internet, as their first venture into cyberspace. In 2015, most real estate companies in Australia had an online presence of one form or another. These online platforms also allow the real estate companies to expand into regional and global real estate markets (Rogers 2016a). The Australian-backed but Asian-based real estate tech start-up Juwai (2014) is a good regional example, and within four years they became one of the largest international real estate websites in the world.

The rise of big real estate data in the early 2000s expanded the scope for uploading real estate, and it was accompanied with the emergence of a suite of big real estate data companies. The Real Estate Transaction Standard (RETS), a real estate data exchange protocol for real estate professionals, was launched in the United States in 1999. This was followed in the early 2000s with the Internet Data Exchange (IDX), a real estate property search site that allowed the public to conduct real estate searches. These types of data companies are beginning to trade in real estate data. In the Asia-Pacific, in 2006, the Korean government introduced the Real Estate Trade Management System to collect real estate transaction data. In 2011, the North American big real estate data analytics company CoreLogic (2014) acquired RP Data, which provides real estate analytical services in Australia and New Zealand. CoreLogic's stated intent was to further expand in the Asia-Pacific region. In 2014, to take a trans-Pacific case, REA Group (2014, 2), which is majority owned by News Corp Australia, a subsidiary of News Corp, announced their 'intention to acquire a 20% stake in Move' (i.e. the first publicly listed internet real estate company mentioned above). According to REA Group (2014, 2), 'News Corp, parent of our majority shareholder, intends to hold the remaining 80%'.

A study of foreign real estate investment looked at the emergence of digital real estate platforms over several decades. It shows that these technologies have been increasingly upscaled in three keys ways: '(1) geographically, at first regionally and then globally; (2) electronically, to include more third-party big data analysis; and (3) socio-economically, to increasingly target, and at times exclusively, high-net-worth individuals and global real estate investment' (see Rogers 2016a, 9). The long-term empirical question that remains for housing scholars with an interest in information technologies is: are these real estate industries shifting their digital technics onto the newer Web 2.0 software paradigms? The move from Web 1.0 to Web 2.0 represents a paradigmatic shift towards big data and big 'data factories' (Burrows and Savage 2014; Keen 2015). Technology writers have grouped the companies in this category under the technological neologism Googlenomics (Keen 2015). These companies give away their software tools and services for free or close to free, but simultaneously become big data companies that 'target their users' behaviour and taste through the collection of their "data exhaust"' (Keen 2015, 60). The 'laws' of Googlenomics proclaim that tech companies can create their own markets by operating in the space between 'the browser', 'search engine and destination content server, as an enabler or middleman between the user and his or her online experience' (Keen 2015, 59).

The history of the internet shows that the danger of Web 2.0 technics is that they drive towards business monopoly and social inequality. Indeed, the biggest Web 2.0 tech companies are banding together to lobby governments, citizenries and even themselves for increasing 'freedom' from government regulation across various digital, economic and socio-political spheres (see The Internet Association 2015). Zeynep Tufekci (2014, 1) even argues that Web 2.0 big data companies 'now have new tools and *stealth* methods to quietly model our personality, our vulnerabilities, identify our networks, and effectively nudge and shape our ideas, desires and dreams' (original emphasis). Their digital practices, according to Tufekci (2014), Isin and Ruppert (2015) and Keen (2015), can be designed to change the way the users of the technologies think about the world and themselves. Tech industry libertarians, following Barlow's (1996, 1) declaration, go further still to argue that cyberspace is the new 'home of mind', and these are not idle threats. The Web 2.0 giant Facebook has conducted secretive online experiments with users' data in an attempt to control their mood, leaving Tufeki (2014, 1) to state that the 'question is not whether people are trying to manipulate your experience and behaviour, but whether they're trying to manipulate you in a way that aligns with or contradicts your own best interests'.

There is some evidence that the larger, more globally focused real estate tech companies are moving towards quasi-Web 2.0 technologies (Rogers 2016a). Unlike the Web 1.0 technologies, these Web 2.0-style companies manage and own the real estate tech products they build. Their tech products include real-time real estate data user interfaces that are typical of the Web 2.0 software paradigm. Some of the real estate tech start-ups have Australian interests, such as the two Australians that founded Juwai (www.juwai.com). Others are right on Australia's doorstep, such as the Singaporean real estate start-up StreetSine Technology Group (www.srx.com.sg). By focusing on these two examples, the next section discusses these types of real estate tech start-ups to expose the real estate technics and data that are being uploaded onto the internet.

Uploading residential real estate

The simplest internet-enabled technology for uploading real estate data is a Web 1.0 investor-focused real estate sales interface. These interfaces place the investor at the centre of a relatively closed network of small data about real estate at a particular site (Rogers 2016a). Perhaps the most common are the website platforms that are built for and then managed by existing real estate companies. These companies upload sales information about individual properties onto their own website with limited analysis or technological manipulation. An Australian example is the real estate company L.J Hooker, who initially commissioned a tech company to upload their local newspaper and real estate shopfront window advertisements onto a web-based platform (ljhooker.com.au). From the early 2000s these Web 1.0 real estate technologies have diversified into third-party real estate websites. In Australia several large news media companies commissioned or secured an ownership stake in the most popular sites, including domain.com.au (Fairfax Media) and realestate.com.au (News Corp, 60 per cent ownership). These real estate technologies allow independent real estate companies and professionals to upload their sales information to a third-party website. It is the technological interface, the 'embodied' and 'situated' experience of the real estate investor in cyberspace, which represents 'a complex interplay between real [i.e. material] and digital geographies' (Isin and Ruppert 2015, 32), that I want to focus on here. For analytical purposes, I have called the digital technic of targeting a real estate investor with a piece of technological hardware and software an *investor-focused* digital act. By using this term I mean to demarcate the technological technic that has been intentionally designed to target a specific population group. The concept of the digital act equally applies, therefore, to real estate professionals (i.e. *professional-focused* digital act) and those who are looking for a rental property (i.e. *tenant-focused* digital act). A good example of a *professional-focused* real estate tech platform in Australia is onthehouse.com.au, which is owned by the Console Group. This real estate tech company, who argue that their 'software has been supporting the Australian real estate industry since 1992', states their 'Console suite' tech product 'is designed to help [real estate professionals] build better relationships, talk to more prospects, increase leads, close more deals, operate with greater efficiency, and monetise your data and website traffic' (Console Group 2015, 1).

The *investor-focused* capabilities of the national real estate websites noted above have more recently been regionally and globally upscaled. In the Asia-Pacific, Juwai claims to operate 'behind China's Firewall [by providing] the most integrated platform connecting international agents and Chinese buyers' (Juwai 2014, 1). The sale of Australian real estate to Chinese nationals through the internet is a key business strategy for this real estate tech company. Juwai (2014, 1) state:

> For Chinese Consumers Juwai.com is an international Chinese platform – hosted in China, entirely in Chinese. Chinese consumers get instant access to international property listings, language and search tools, as well as relevant research and information they need to make informed decisions about overseas property purchasing.

In terms of the small real estate data that is flowing through their tech platform to target Chinese investors, for a fee Juwai translates local Australian (and other countries') dwelling-specific real estate data from

English to Mandarin. They provide in-house cross-cultural and language translations and Chinese social media compatibility: 'Our professional editorial team translates in a style and tone that resonates with Chinese buyers', and Juwai's 'Mobile App with Chinese social channel integration [is] combined with online Chinese social media features' (Juwai 2014, 1).

The more sophisticated Web 2.0-style *professional-focused* real estate sales technologies place real estate and other professionals at the centre of a diffuse network of big and small data about investors, property developers, immigration agents, financial institutions and other real estate information (Rogers 2016a). Within the real estate tech industry, Web 1.0 and quasi-Web 2.0 platforms can operate independently or they can sit happily alongside each other within a broader tech platform. Juwai's portable online analytical tool for real estate professionals is a good example, and it operates alongside their Chinese *investor-focused* digital acts. Their *professional-focused* real estate platform networks real estate professionals from around the world into 'Juwai's exclusive audience of 1.5 million high-net-worth Chinese consumers' (Juwai 2014). The global real estate company Engel & Völkers (2014) also developed an online real estate *professional-focused* product called 'my life', which they describe as 'Practical Knowledge for Sales Advisors'. StreetSine (2014), in Singapore, also 'integrate big data sets with mobile workflow applications to help real estate-related organisations and professionals employ real-time, relevant, proprietary information in the marketing of their products and services'. Their tablet-friendly platforms 'provide the property market with computer-generated pricing', which is a tech product they have trademarked as Home Report™. StreetSine (2014) market their Singapore Real Estate Exchange platform, trademarked as SRX™, to 'property-related professionals', including real estate agents, bankers and lawyers. Much like Juwai, their *professional-focused* digital acts operate alongside a suite of *investor-* and *tenant-focused* digital acts, with their tech products also targeting real estate buyers, sellers, landlords and rental tenants. StreetSine's big real estate data set can provide extremely fine-grained information about Singaporean real estate, all the way down to a 'computer-generated price' for an individual dwelling. They call this computer-generated price an X-Value™. As the real estate professionals use SRX™, the technology captures their real estate data exhaust, thereby growing the company's data set. However, the real estate data set that produces the X-Value™ is only available because of the government's historic role in the provision of public housing in Singapore (Chua 1997). The government collects the bulk of the real estate data that this technology is built upon. Thus, the role of the government as a real estate data collector and provider is central to StreetSine's Web 2.0-style real estate technology.

Isin and Ruppert (2015, 26) argue that these new socio-technical relationships enable a special class of digital technic and 'speech act', and what is at stake in these digital technics and acts is 'the production, dissemination, and legitimation of knowledge'. More recently, the Web 1.0-

and Web 2.0-style real estate platforms have had multimedia channels incorporated into the technologies to broadcast high-quality digital audio/visual discursive content about the private real estate market. StreetSine has a regular national radio interview and podcast slot on Singapore's ONE FM radio station. The podcast program covers real estate topics such as real estate prices, home loans and the local private property market. They have been critical of the Singaporean government's affordable housing interventions, such as the foreign investment 'cooling measures' that attempted to mitigate house price increases (e.g. 'Condo market hit hard by cooling measures' podcast on www.srx.com.sg/podcasts). A real estate tech start-up executive I interviewed in 2014 stated: 'now I'm a free-market capitalist-libertarian … I can see how information technology can help … allow this asset class [residential real estate] to be a more frequently traded asset class'. The discursive content within these types of multimedia tech products and the statement by the real estate tech executive show that a free market liberalism, which is built upon private property ideologies and underwritten by a disdain for government intervention, is deeply embedded within some of these new real estate technics and digital acts.

The recent formation of the political lobby group The Internet Association (2015) is a good pan-tech industry example. This group is an alliance of some of the largest and wealthiest Web 2.0 tech companies, including Facebook, Google and Amazon. This lobby group argues that every tech company should be 'uninhibited' by government taxation, regulation and censorship. They also argue that they should not be held responsible for the user-generated digital acts they enable through their platforms. For example, they state that 'Internet intermediaries must not be held liable for the speech and activity of Internet users' (The Internet Association 2015, 1). The Internet Association is advocating for an instrumental computer science view of technology, which allows the lobby group to position the 'freedom' to produce electronic code in cyberspace as more important than thinking about an appropriate regulatory environment that might be used to guide their digital acts, and the resultant effects in the material world (Burrows and Savage 2014; Isin and Ruppert 2015; Keen 2015).

The practice of uploading real estate onto the internet is not developed in a political vacuum, and there are many examples of resident-led or researcher-led tech resistance to market-centric real estate practice. It is not my intention to explore these alternative forms of digital real estate technics here. However, The Anti-Eviction Mapping Project (United States) and Our House Swap (Australia) are two exemplary cases that are worthy of mention. The Anti-Eviction Mapping Project is a web-based 'data analysis collective' that uses a crowd-sourcing platform to recruit geographic information system (GIS) specialists to document 'the dispossession of San Francisco Bay Area residents in the wake of the Tech Boom [Web] 2.0' (Anti-Eviction Mapping Project 2015, 1). The public housing tenant-managed Our House Swap website bypasses the state housing authorities' role in mediating tenants' residential mobility choices (Our House Swap 2015, 1). These types of digital initiatives are enframed by anti-gentrification and collective ownership ideologies that represent counter-discourses to market-centric forms of real estate technic. The Anti-Eviction Mapping Project, Our House Swap, StreetSine and Juwai website technologies show that underneath these technological platforms are a set of organising real estate technics and housing ideologies, which have quite literally been uploaded as forms of real estate or housing power and/or resistance.

The real estate technics that are now associated with buying and selling real estate as private property in the globalising digital real estate industry have a clear lineage back to pre-internet real estate technics in countries like Australia (Rogers 2016b). Over the last century, the real estate industry has categorised their real estate technics and ascribed to their many real estate actions a set of linguistic referents, which are: (1) capital costs; (2) capital lending; (3) growth; (4) yield; (5) liquidity; (6) transaction costs; (7) taxation costs; and (8) ownership rules (Rogers 2016a; Tiwari and White 2010). These technic referents signify the broader set of actions that the real estate professionals, investors, tenants, governments, financial institutions and other actors come together to undertake. I have tracked the repeated use of these real estate technics back in time, to show how they became normalised and turned into conscious and transferable real estate technics (see Rogers 2016b). The agonistic struggles that underwrite these types of performative technics, as Isin and Ruppert (2015, 33) point out, transcend the material/ digital world divide – the digital acts are always *in the world*. The pre-internet language acts and real estate technics of local real estate professionals, such as real estate advertisements in local newspapers and shopfronts, were readily uploaded onto the internet as regional and global technics. Therefore, there is a larger historical context that frames the emergence of internet-enabled real estate technologies, which includes the transfer of market-centric real estate technics from pre-internet into internet-enabled real estate technologies. The historical mapping of these real estate technics could further expose how these real estate technologies are mediated across different technological, generational and geographical scales (Rogers 2016b). To understand twenty-first-century digital real estate technologies, housing scholars need to know more about how real estate technics are transferred from one mediating technological to another, from one generation of real estate professional to another, and from one investor to another.

Conclusion

A global digital expansion of the residential real estate industry is underway. Millions of local residential homes from some of the wealthiest countries around the world are being uploaded for sale onto thousands of internet-enabled real estate platforms. While critical housing scholars (Crabtree 2013) and some digital citizenries (Our House Swap) are abandoning and thinking beyond the technics of housing and land as private property, the globalising real estate industry is digitising, uploading and upscaling the local real estate technics of a former era onto their new tech products. In the Asia-Pacific, the real estate tech industry has uploaded the real estate technics that built the Great Australian Dream. James Kirby (2002) jokes in his financial self-help book *Investing for Dummies* (which, as a speech act, is an important pre-internet real estate mediating technology), 'In Australia ... property is like a game and is followed like the footy'. However, this globalising real estate game is no joke and it has already produced significant housing inequity is several Australian cities. The real estate tech entrepreneurs are uploading these local real estate technics, which created significant housing inequity and exacerbated housing disadvantage in the second half of the twentieth century, as international real estate technics for the twenty-first. These globalising real estate technologies are not presently spurring on creative innovation for a more democratic technics of housing and land. They are not testing new technics that might address the increasingly unaffordable housing landscapes in many global cities around the world. They are not using technology to imagine or discover innovative ways to retool real estate and housing so that it is more equitable and freedom-creating, which are the ideals that the internet was supposedly built upon.

The new real estate technologies that host these market-centric real estate technics readily account for the legal frameworks of different nation-states and the cultural and linguistic barriers that previously restricted transcultural and transnational real estate sales. These real estate technology entrepreneurs are building tech products that have private property and market-centric ideologies built into their very functionality. A further task for housing scholars is, therefore, to analyse the way the real estate investors and professionals are 'downloading' this information onto their smartphones and tablets; to think through the relationships between the user, the technological object and the information that is flowing through a given embodied user and technological object assemblage. The conceptual framing of such a project needs to deal with the way the users interact with these new types of real estate platforms and data to ask whether we can record the subjectivities, if any, that result from these new digital real estate technologies.

Finally, the rise of Web 2.0-style real estate technologies, and the buying and selling of real estate tech companies, has made real estate data *itself* a sought after and tradable commodity. Government regulation should ensure that big real estate data become open source, publicly available and free. Internet history shows that large tech companies are not committed to free open-source big data capture, transparency and social democracy, and the large Web 2.0 tech companies are driving towards technological oligarchy in their respective entrepreneurial fields. What might be lost if real estate tech companies follow suit is the ability to creatively innovate outside the sets of market-centric real estate technics that framed twentieth-century real estate practice. In short, without government intervention the online market-centric real estate technics of the twenty-first century, much like the local offline technics of the twentieth century, could negatively affect global housing equity.

References

Anti-Eviction Mapping 2015. *Anti-Eviction Project*. Retrieved on 2 June 2015 from www.antievictionmappingproject.net.

Barlow, J.P. 1996. *A Declaration of the Independence of Cyberspace*. Retrieved on 4 June 2015 from https://projects.eff.org/~barlow/Declaration-Final.html.

BrickX. 2015. *BrickX*. Retrieved on 8 June 2015 from www.brickx.com.

Burrows, R. and Savage, M. 2014. After the crisis? Big Data and the methodological challenges of empirical sociology. *Big Data & Society* 1(1): 1–6.

Chua, B.-H. 1997. *Political Legitimacy and Housing: Singapore's Stakeholder Society*. Routledge: Abingdon.

Console Group. 2015. *Solutions for Real Estate Agents*. Retrieved on 12 July 2015 from www.onthehousegroup.com.au.

Corelogic. 2014. *Corelogic Announces Completeion of RP Data Acquisition*. Retrieved on 12 July 2015 from www.corelogic.com/about-us/news/corelogic-announces-completion-of-rp-data-acquisition.aspx

Crabtree, L. 2013. Decolonising property: exploring ethics, land and time through housing interventions in contemporary Australia. *Environment and Planning D: Society and Space* 31: 99–115.

Dufty-Jones, R. 2016. Housing and home: objects and technologies of neoliberal governmentalities. In S. Springer, K. Birch and J. MacLeavy (eds), *Handbook of Neoliberalism*. Routledge, London.

Ellul, J. 1954. *The Technological Society*. Vintage Books, Toronto.

Ellul, J. 1981. *A Temps et à Contretemps*. Centurion, Paris.

Engel & Völkers. 2014. *Practical Knowledge for Sales Advisors*. Retrieved 28 June 2015 from www.engelvoelkers.com/company/academy-further-training-education/academy-video-my-life.

Ferguson, N. 2008. *The Ascent of Money: A Financial History of the World*. Penguin, London.

Franks, R. and Cook, P. 1996. *The Winner-Take-All Society: Why the Few at the Top Get So Much More than the Rest of Us*. Penguin, New York.

Heidegger, M. 1977 [2013]. *The Question Concerning Technology and Other Essays*, trans. W. Lovitt. Harper Perennial, London.

The Internet Association. 2015. *We are the United Voice of the Internet Economy*. Retrieved on 20 July 2015 from http://internetassociation.org.

Isin, E. and Ruppert, E. 2015. *Being Digital Citizens*. Rowman & Littlefield, London.

Juwai. 2014. *Juwai*. Retrieved on 12 March 2015 from list.juwai.com

Keen, A. 2015. *The Internet is Not the Answer*. Atlantic Books, London.

Kirby, J. 2002. *Investing for Dummies: Identify and Manage the Best Investment for You*. Wiley Publishing Australia, Milton.

Kittler, F. 1995. *Discourse Networks 1800/1900*. Stanford University Press, Stanford.

Kittler, F. 2002. *Optical Media*. Polity Press, Cambridge.

Lovitt, W. 2013. Introduction. In M. Heidegger, *The Question Concerning Technology and Other Essays*. Harper Perennial, London.

Manyika, J., Chui, M. Brown, B. Bughin, J. Dobbs, R. Roxburgh, C. and Hung Byers, A. 2011. *Big Data: The Next Frontier for Innovation, Competition, and Productivity*. McKinsey & Company, New York.

Mazzucato, M. 2013. *The Entrepreneurial State: Debunking Public vs. Private Sector Myths*. Anthem Press, New York.

Millward, S. 2014. *Right Place at the Right Yime: How Two Australians Created China's Most Perfect Startup*. Retrieved on 29 August 2015 from www.techinasia. com/story-juwai-overseas-property-portal-china.

Mitcham, C. 1994. *Thinking Through Technology*. University of Chicago Press, Chicago.

Movoto. 2014. *Real Estate Through Time Infographic*. Retrieved on 20 April 2015 from www.movoto.com/blog/infographic/real-estate-through-time-infographic.

Mumford, L. 1934. *Technics and Civilization*. Harcourt, Brace and Company, New York.

Ortega, J.-y.-G. 1939. *Meditación de la Técnica*. Retrieved on 24 March 2015 from http://monoskop.org/images/d/d4/Ortega_y_Gasset_Jose_1939_1964_ Meditacion_de_la_tecnica.pdf.

Our House Swap. 2015. *Our House Swap*. Retrieved on 7 June 2015 from www. ourhouseswap.com.au.

REA Group. 2014. *Annual Report 2014*. Retrieved on 15 June 2015 from www. rea-group.com/irm/content/ar2014/annualreport.pdf.

Rogers, D. 2016a. Becoming a super-rich foreign real estate investor: globalizing real estate data, publications and events. In R. Forrest, D. Wissink and S. Koh (eds), *Cities and the Super-Rich: Real Estate, Elite Practices and Urban Political Economies*. Palgrave Macmillan, Basingstoke.

Rogers, D. 2016b. *The Geopolitics of Real Estate: Reconfiguring Property, Capital and Rights*. Rowman & Littlefield, London.

Rogers, D., Lee, C.L. and Yan, D. 2015. The politics of foreign investment in Australian housing: Chinese investors, translocal sales agents and local resistance. *Housing Studies, i-First*. DOI: 10.1080/02673037.2015.1006185

Rousseau, J. 1750. *Discourse on the Sciences and Arts*. Noël-Jacques Pissot, Paris.

Smith, S. 2008. Owner-occupation: at home with a hybrid of money and materials. *Environment and Planning A* 40(3): 520–535.

StreetSine Technology Group. 2014. *StreetSine Technology Group*. Retrieved on 24 July 2015 from www.streetsine.com.

Tiwari, P. and White, M. 2010. *International Real Estate Economics*. Palgrave Macmillan, Basingstoke.

Tufeki, Z. 2014. *Facebook and engineering the public*. Retrieved on 15 March 2015 from https://medium.com/message/engineering-the-public-289c91390225.

Winthrop-Young, G. 2011. *Kittler and the Media*. Polity Press, Cambridge.

Zipreality. 2014. *Zipreality*. Retrieved on 26 May 2015 from www.ziprealty.com/ about_zip/index.jsp.

3 Homeownership, asset-based welfare and the actuarial subject

Exploring the dynamics of ageing and homeownership in New Zealand

Laurence Murphy and Michael Rehm

Introduction

Since the mid-twentieth century, homeownership has assumed a dominant position within certain Anglophone societies (Smith 2015). The rising popularity of homeownership has been variously ascribed to a combination of individual consumer preferences, reflecting an ongoing search for 'ontological security' (Saunders 1990), and government housing policies that have supported the expansion of the tenure (Roland 2008). From the early 2000s, within the context of significant mortgage market liberalisation at an international level, homeownership in countries with mature mortgage markets has been increasingly viewed as a financial asset (Smith and Searle 2010). The twin attributes of housing as a commodity and an asset sit at the heart of a set of dynamics that are re-casting popular and political discourses surrounding the nature and future of homeownership. In contrast to the relative fixity of housing as a physical entity (consisting of bricks and mortar, or weatherboard and tin roofs), the socio-economic characteristics and experience of homeownership as a tenure are subject to change. Smith (2008, 521), in examining the 'complex, politically charged, and ethically challenging entanglements between the materiality of housing, the meaning of home, and the mobilisation of money' alerts us to political and material processes that are re-positioning homeowners from 'property holding citizens' to 'asset accumulating investors'. Moreover, her analysis engages with the 'grammars of living that differentiate among homebuyers' in a sector that 'accommodates a very wide range of households' (Smith 2008, 525). In this chapter we explore the changing nature of home as an assemblage of money, materials, ideas and practices, while explicitly recognising that homeowners experience housing in a complex manner that reflects a set of socio-economic factors and disparate geographies.

At a general level, Australia and New Zealand share a common homeownership culture. Owner occupation is the dominant tenure in both countries and from the 1950s onwards, Castles (1996) argues, governments in Australia and New Zealand viewed homeownership as a welfare good and supported its expansion. The detached single dwelling has traditionally dominated both housing markets, and household wealth is increasingly tied to the performance of housing markets (Ong *et al.* 2013a, 2013b; Smith 2015). Significantly, since the mid-1980s, the top four mortgage banks in New Zealand are Australian-owned. Yet, in both countries the nature and experience of homeownership is changing. Housing equity withdrawal has made housing assets more fungible and altered homeowner consumption practices (Ong *et al.* 2013a, 2013b; Parkinson *et al.* 2009). Rising house prices are challenging the Australian and New Zealand dream of owning a home, but homeownership continues to be constructed in public and policy discourses as a suitable vehicle for wealth accumulation. Owning or not owning your home has significant financial and material implications for ageing Australians (Colic-Peisker *et al.* 2015) and New Zealanders. Moreover, as the dramatic experience of the US subprime mortgage market crisis has shown (Aalbers 2012; Immergluck 2011), the dominant role of homeownership in Australia and New Zealand carries with it significant individual (homeowner), institutional (banks and financial institutions) and public (state) risks (Murphy 2011).

In this chapter we examine a set of dynamics that are reshaping the perception and experience of homeownership in housing systems 'anchored on owner occupation and financed by mortgage debt' (Smith 2015, 62). Focusing on issues of risk, wealth accumulation/decumulation and 'emergent' housing/welfare policy, we explore a set of evolving tensions in individual and governmental understandings of the nature and function of homeownership. In particular, we chart the changing relationship between homeownership and an ageing population. Where, once, homeownership was constructed as a 'housing career' destination, increasingly it is being viewed as a store of wealth that can/should be mobilised to meet the welfare needs of old age (Ong *et al.* 2013a, 2013b). In this chapter we argue that the rise of 'asset based welfare' ideas (Murphy 2012) that construct housing as a pension are based on a particular interpretation of homeownership that elides the realities of a tenure that is highly differentiated. Moreover, the evolving policy discourse underpinning the notion of housing as a pension increasingly positions homeowners as a type of risk analyst, or actuary, engaged in strategic investments to secure their personal welfare in old age.

This chapter offers an extended critical review of a set of literatures relating to homeownership and asset-based welfare policies. In particular, we chart the rise of asset-based welfare ideas and their links to policy concerns relating to ageing societies and homeownership. In addition, we review the emerging policy dynamics that are helping to reshape policy practice and the experience of homeownership. Drawing on the preliminary findings of research into ageing and downsizing in New Zealand, we offer a brief empirical analysis of the differentiated realities of ageing, homeownership and housing wealth accumulation. We argue that amid the changing ideology of homeownership and reformulation of policy imperatives, the lived experiences of homeowners are producing complex, spatially differentiated outcomes that challenge the logic of asset-based welfare policies. While our empirical analysis is New Zealand-focused, our argument has significant relevance for understanding contemporary processes reshaping the ideological and material experiences of older homeowners in Australia (Judd *et al.* 2014; Ong *et al.* 2013a, 2013b). Moreover, our analysis highlights that while housing policy in countries with mature mortgage markets increasingly recognises that home is a product of an array of financial interactions, questions of social equity, resident capacities and distributional justice are as important as ever.

Shifting understandings of homeownership

Increasingly, it is recognised that housing markets are made, not given. Consequently, markets can be viewed as assemblages, defined as 'complex entanglements or networks of humans, materials, institutions, politics and technologies' (Lovell and Smith 2010, 458). In this context, the rise of homeownership reflects the operation of a set of legal, financial, material and discursive practices. Moreover, technologies of governance have increasingly positioned homeownership in terms of 'responsible citizenship' and 'ethical living' (Smith 2015).

Homeownership has long been connected with political discourses that associate the tenure with a range of values including 'self-esteem', 'responsibility' and citizenship'. Forrest and Hirayama (2015) position political support for homeownership from the 1950s through to the 1970s as part of the 'era of "embedded liberalism" or post-war Keynesianism' (p. 234). In this period, homeownership was increasingly a central component to housing policies that promised the fulfilment of localised versions of the 'American Dream'. Significantly, during this phase of expansion homeownership was imbued with a set of specific meanings that Fox O'Mahony and Overton (2015) argue have shaped older owners' understandings of the tenure. They argue that official discourses centred on the 'putative benefits of ownership as the vehicle *par excellence* to achieving a meaningful connection with *home*' and mobilised a 'rhetoric of independence, control, shelter, security, steady saving and capital asset' (p. 394).

Arguably, the expansion of homeownership in the post-war Keynesian period constituted something of a golden era in its history. Homeowners achieved access to the middle class and enjoyed a range of material benefits (Saunders 1990). However, this expansion necessarily altered the socio-demographic character of the sector and individual experiences of the tenure. Despite a rhetoric that constructs homeownership as a homogeneous entity, the reality is a tenure characterised by differentiation and fragmentation (Forrest *et al.* 1990). House price appreciation varies temporally and spatially and alters the potential for housing to act as a source of wealth accumulation (Malpass 2008). Moreover, the position of marginal homeowners in the tenure is often precarious and, as the subprime mortgage crisis so forcefully demonstrated, mortgage arrears and foreclosures can lead to people exiting the tenure under considerable financial stress (Aalbers 2012; Immergluck 2011). In Australia Colic-Peiskar *et al.* (2015) have highlighted the problems of older households who have dropped out of homeownership, while Ong *et al.* (2013b) identify the complex and problematic nature and implications of housing equity withdrawal for older households.

The advent of neoliberalism, with its emphasis on deregulation and private market provision, underpinned the dramatic changes that have characterised global mortgage markets since the 1990s (Gotham 2009). In addition to reshaping housing markets, the rise of neoliberalism profoundly altered the nature and understandings of the welfare state. Of particular relevance to our analysis are the policy responses that have been emerging to address the consequences of ageing societies (Murphy 2012). Under traditional welfare state practices, based on income support, an ageing society combined with higher dependency ratios exposes the state to a potential fiscal blowout. To address the potential of a growing public pension 'burden', governments around the world have been actively promoting private and compulsory superannuation schemes. In effect, they have been refashioning welfare systems in a manner that positions individuals as 'responsible citizens' that need to take care of their future welfare needs (Murphy 2012). One influential policy discourse that has emerged centres on the idea of asset-based welfare. At the core of asset-based welfare is the idea that policies should move away from the traditional concern of income support and move to encourage the accumulation of assets that can be mobilised to meet future social needs (Sherraden 2003). Significantly, the logic of asset-based welfare aligns with long-held government support for homeownership. Under this new welfare agenda, homeownership is increasingly being viewed as a 'pension' (Doling and Ronald 2010; Groves *et al.* 2007; Malpass 2008; Ong *et al.* 2013a, 2013b).

The realignment of homeownership as a central component of a new welfare state is problematic (Malpass 2008). Drawing on a review of the literature, Murphy (2012) identifies several limitations of the homeownership/asset-based welfare argument. First, the price performance of housing markets is highly uneven and, in general, housing markets tend to exacerbate wealth inequalities. Second, housing markets are usually in decline during periods of greatest social welfare need. Thus, it is likely that rising levels of mortgage default, repossessions and negative equity will coincide with periods of greatest social need. Third, the logic of mobilising housing wealth to meet social needs in old age is challenged by cultural practices and expectations, and by emerging trends of housing equity withdrawal (HEW) among younger homeowners. Cultural practices on the part of older homeowners can lead to resistance to using the home as a pension (Fox

O'Mahony and Overton 2015) and rising HEW among younger cohorts of homeowners may result in an increasing number of households retiring with mortgage debt (Smith and Searle 2010), thus reducing the amount of wealth available to meet welfare needs in old age.

Notwithstanding the problems associated with re-positioning homeownership within a 'new welfare state', it is clear that homeownership is implicated in emerging wealth inequalities and that homeowners are actively engaged in equity withdrawal processes (Parkinson *et al.* 2009; Smith and Searle 2010). Moreover, older homeowners, who have secured housing equity, are increasingly considering their future housing and welfare needs (Ong *et al.* 2013b). For those who opt for 'ageing in place', new financial products can be mobilised to release equity to meet consumption needs (Fox O'Mahony and Overton 2015). Alternatively, older owners are increasingly considering the potential for downsizing, a process in which as a result of moving home they release housing equity (Judd *et al.* 2014). In addition, governments around the world are increasingly introducing asset-testing criteria in relation to rest home subsidies. These policies usually require homeowners to release equity, or sell the family home, to cover residential care costs (Murphy 2012). Thus, in terms of individual homeowner practices and government policies, the experience and ideology of homeownership is being constantly refashioned. In the next section we reflect on the nature of these changes.

Homeownership, asset-based welfare and the actuarial subject

Attempts to reposition homeownership within an emergent asset-based welfare system rest on a set of ideological transformations. Underpinning these transformations was the emergence of neoliberal economic and political processes that emphasised financial liberalisation and global flows of capital (Forrest and Hirayama 2015). The traditional ideological and political logics that propelled the growth of homeownership were subject to considerable change under neoliberalism. For the purpose of developing our argument we will address two key issues. First, we explore the manner in which neoliberalism altered the practices of homeowners, paying particular attention to the ways in which the liberalisation of mortgage markets altered homeowners' willingness to engage in equity withdrawal practices. Second, we comment on how the discourse of housing as an engine of wealth creation has emerged in political understandings of the nature of homeownership. We argue that the intersection of these processes has particular implications for emerging understandings of home, homeownership and an ageing society. In particular, we argue that the social dimensions of homeownership and its links to citizenship (Forrest and Hirayama 2015) have given way to a political process in which homeowners are constructed as an 'actuarial subject' (Price and Livsey 2013).

During the era of 'embedded liberalism' (Forrest and Hirayama 2015), the expansion of home ownership was associated with a rhetoric and discourse that emphasised notions of citizenship and financial security, and resulted in 'policy-led associations between ownership and a meaningful attachment to home [that] generated powerful "discourses of normalisation"' (Fox O'Mahony and Overton 2015, 395). Homeownership was aligned with a set of affective outcomes which emphasised the psychological and social dimensions of home, as well as the investment dimensions of owning a property. Reviewing a range of research, Fox O'Mahony and Overton (2015) highlight the multiple meanings that older homeowners invest in their homes. They argue that 'the symbolic meaning of home for older people reflect its status as a physical representation … of the life-course and its achievements and events … its function as a stable base for the preservation of identity for the ageing self' (Fox O'Mahony and Overton 2015, 395). Moreover, they argue that paying off the mortgage (unencumbered homeownership) was central to notions of financial security for older people.

The rise of neoliberalism, with its emphasis on privatisation, market liberalisation and self-reliance, has altered the narrative and practice of homeownership. Under conditions of significant mortgage market liberalisation, the home has increasingly assumed the characteristics of an investment vehicle (Smith 2015). In countries such as the United States, United Kingdom, Australia and New Zealand, equity withdrawal became prevalent and the home was increasingly enmeshed in discourses that positioned housing as a new ATM (Parkinson *et al.* 2009; Smith and Searle 2010). 'Cash-out refinancing' and 'home equity line of credit' loans became important mechanisms for homeowners to extract equity and, at the same time, increase their levels of leverage and potential risk (Immergluck 2011). Significantly, the purpose of this equity withdrawal was often not about conspicuous consumption, but centred on the welfare needs of households. In particular, Searle and Smith (2010) and Parkinson *et al.* (2009) show how households used equity withdrawal as a form of insurance to meet household budget crises arising from biographical events such as the birth of a child, unemployment or relationship breakups. Notwithstanding its purpose, the rise of equity withdrawal marked an important moment in a transition 'away from the idea that housing wealth is primarily a legacy for future generations towards the notion that it is a resource to spend across the life course' (Smith *et al.* 2008, 89).

The increasing level of household wealth tied up in housing and the rising fungibility of housing as an asset has coincided with an ongoing political debate concerning the role of the welfare state in an ageing society. The prospect of an ageing society is increasingly constructed as problematic for welfare systems where public pensions are funded by taxing those who are currently in employment. Across the OECD, governments are grappling with the potential fiscal consequences of demographic change. In the UK it has been estimated that an ageing demographic will add £80 billion to pension and healthcare costs and that expenditure on long-term rest care will increase substantially (Searle and McCollum 2014). In light of the potential increased demands on the welfare state it is unsurprising that the wealth tied up in housing has increasingly been viewed as a welfare resource (Murphy 2012; Ong *et al.* 2013a, 2013b). In Great Britain it has been estimated that the total net property wealth for all homeowners is £3,375 billion and the median household property wealth is £148,000 (Searle and McCollum 2014). Significantly, while debates over the nature of pension provision have raged for decades, Searle and McCollum (2014) identify an important shift in the discourses surrounding ageing. They identify a shifting political rhetoric in which 'older people have moved from being poor and deserving to being the hoarders of societies' wealth and welfare' (Searle and McCollum 2014, 326).

At the centre of this emerging discourse is the role of housing wealth. Within the emerging logic of asset-based welfare policy developments, the wealth accumulated in housing should be mobilised to meet individual welfare needs. In order to facilitate this, a new ideology of homeownership is emerging that challenges the earlier constructions centred on 'security', 'saving' and an 'unencumbered' homeownership in old age. Fox O'Mahony and Overton (2015, 395) argue that the narrative of an asset-based system is founded on a model of homeownership whereby the value of the asset is spent through 'practices of accumulation and decumulation enabling the personal achievement of self-responsible neo-liberal citizenship'.

The evolving policy discourses around financing later life are predicated upon the availability of appropriate financial products (private pensions, mortgage products) and a financially sophisticated citizenry. Price and Livsey (2013) offer an insightful and critical evaluation of this emerging policy dynamic in their analysis of the financing of pensions and long-term care. In particular, they highlight the manner in which the asset-based welfare rhetoric constructs individuals as 'actuarial subjects'. In reflecting on the policy imaginary underpinning a welfare system based on individual responsibility, they argue that citizens are 'constructed as malleable subjects who can be "nudged" and educated to become financially capable, fiscally competent, actuarially aware subjects' (Price and Livsey 2013, 67). The term 'actuarial subject' adds to the notion of the neoliberal (self-governing) subject by placing emphasis on a notional or actual set of 'calculative practices' that are undertaken by investors rather than citizens. It points to a form of agency that individuals may, or may not, be able to exercise.

Price and Livsey (2013) trace the UK state's transition from being a 'universal provider' in welfare to it being an 'enabler'. Under this policy trajectory, while individuals are constructed as 'actuarial subjects' assessing their lifetime risks and finances, their choices and financial outcomes are determined by processes that are often beyond their capacity to manage. In reflecting on pension and housing dynamics, Price and Livsey (2013, 81) alert us to the importance of geography and argue that securing housing equity is 'mostly a geographical accident of where and when they purchased their home'.

From our review of the literature we argue that two key processes are at work. First, at an ideological level, neoliberal economic and welfare dynamics are recasting homeownership within a new set of financial imperatives and welfare provision processes (Malpass 2008) centred on the notion of the 'actuarial subject' (Price and Livsey 2013). This ideological shift (Fox O'Mahony and Overton 2015; Searle and McCollum 2014) rests on a generic understanding of the benefits of homeownership that necessarily ignores the spatio-temporal dynamics that shape the lived realities of homeowners, and older households that might drop out of home ownership (Colic-Peisker *et al.* 2015). Moreover, the new ideology of homeownership clashes with the rationales that propelled the growth of this tenure and this is particularly true for older homeowners (Fox O'Mahony and Overton 2015). Second, in understanding the role of homeownership in welfare provision it is necessary to take account of the fragmented and differentiated nature of the tenure (Murphy 2012). To explore some of these issues in more detail, we now turn to an analysis of housing dynamics in New Zealand.

New Zealand, ageing and homeownership

Australia and New Zealand have been fully implicated in the ideological and material refashioning of home and homeownership under neoliberalism. In Australia, Parkinson *et al.* (2009) note equity borrowing occurring earlier in the life-cycle than expected under more traditional practices that had emphasised housing wealth as something to be passed on as an inheritance. Moreover, they raise concerns regarding the prospect of homeowners reaching retirement with elevated debt levels. Ong *et al.* (2013b) show that older homeowners in Australia are taking on more equity debt prior to retirement, and for those above pension age there is a shift to 'downsizing and selling up' as the key forms of equity withdrawal. Significantly, arising from their modelling, Ong *et al.* (2013b, 84) argue that 'policies designed to encourage older home owners to tap into their housing wealth beyond current average amounts of HEW to say, fund aged care needs, may expose many to undesirable levels of limited equity risk'. Moreover, their modelling indicates that 'downsizing and selling up' adversely affects means-tested income support programmes and this has significant implications for household welfare.

In this section we turn our attention to processes at work in New Zealand. In particular, from the starting point that home is an assemblage of money and materials (Smith 2008), we are concerned with examining emerging calculative practices that position home as a possible welfare safety net. These calculative practices, centred on measures of average house prices and the value of the total stock of housing, are at work in shaping housing policy. These measures inform and influence policy trajectories but also affect the decision-making environment of ageing homeowners. Moving beyond broad measures of housing market performance, we focus on how the reality of socio-economic diversity and the geography of housing markets disrupt and make problematic policy constructions of homeowners as actuarial subjects. We maintain that the processes that we examine in New Zealand resonate with the issues (homeownership trends, equity withdrawal, welfare policy and ageing) emerging in Australia.

In common with other OECD countries, New Zealand's population is ageing and this has significant implications for welfare policy and government expenditures (New Zealand Treasury 2013; Saville-Smith 2013; St John *et al.* 2012). Of particular importance for policy-makers is the rising number of people aged over 85 years, as it is this group who are likely to make the most demands on health and residential care facilities and government funding. It is estimated that the number of people aged over 85 will increase from 76,000 in 2012, depending on the assumptions, to between 290,000 and 430,000 in 2061 (St John *et al.* 2012). In addition, New Zealand's ageing society will be accompanied by a decline in the number of persons aged 15 to 65 years compared to those over 65 years. This ratio is expected to decline from 5:1 in 2006 to 2:1 by 2050 (New Zealand Treasury 2013), and this has significant implications for the New Zealand superannuation scheme, which is a pay-as-you-go system (St John *et al.* 2012).

Older New Zealanders have a strong reliance on government superannuation as their main income source. Saville-Smith (2013, 8) notes that 'sixty percent of older New Zealanders received between 80 percent to all of their income from New Zealand superannuation and government supplements'. While there are debates concerning the adequacy of older people's income, Saville-Smith (2013) notes the importance of mortgage-free homeownership in maintaining older people's living standards. In this context, it is clear that outright homeownership functions as an important restraint on poverty in old age and, more generally, has the potential to modify pension demands (Murphy 2012). This 'traditional' function of homeownership in New Zealand accords well with the twentieth-century ideology of homeownership identified by Fox O'Mahony and Overton (2015) and aligns with the Australian experience, where 'retirement income policy is built around the assumption that state pensions can be kept low because an overwhelming majority of older Australians are outright homeowners' (Colic-Peisker *et al.* 2015, 167).

The role of housing and housing wealth in New Zealand has increasingly changed since the 1990s. Reflective of new demographic trends and rising house prices, homeownership levels declined from 73.8 per cent in 1991 to 64.8 per cent in 2013 (Statistics New Zealand 2014). Figure 3.1 shows home ownership rates by age for the period 2001 to 2013. Significantly, homeownership rates for younger age groups have declined more dramatically than for older households. These data are important on two levels. First, they indicate that homeownership is increasingly concentrated within older age groups. This process has significant policy import. If outright homeownership is constructed as a means of reducing poverty in

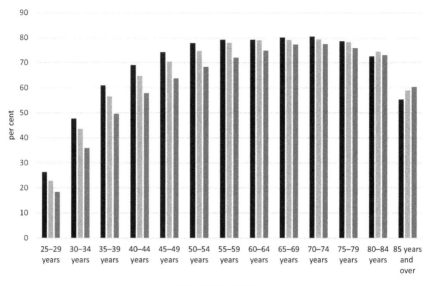

Figure 3.1 Homeownership by age.
Source: 2013 Census QuickStats About Housing: retrieved on 12 May 2015 from www.stats.
govt.nz/Census/2013-census/profile-and-summary-reports/quickstats-about-housing.aspx.

old age (by ageing in place), then there are benefits in keeping people in homeownership. However, if homeownership is viewed as a financial asset then there is the potential for decumulation and equity withdrawal among older households. Second, if the declining rate of homeownership represents a structural shift in the housing system, then the capacity for homeownership to function as a central pillar of an asset-based welfare system in the future is limited.

Notwithstanding the changing dynamics of homeownership rates, rising house prices have had a significant impact on household wealth in New Zealand. As Figure 3.2 shows, between 1990 and 2014 the total value of private dwellings rose from $123 billion to $725 billion. Constructed as a financial asset, the substantial stock of wealth tied up in housing is increasingly viewed, from a policy perspective, as something that households can use to support their welfare needs. In this context it should be noted that residential aged care subsidies have long been subject to asset testing (St John *et al.* 2012). Moreover, in response to a fear of asset testing, an increasing number of households have placed their homes in family trusts and effectively legally alienated themselves from their properties (St John *et al.* 2012). The rise in the number of homes in family trusts, over 215,000 in 2013, points to an interesting paradox at work in the development of asset-based welfare policies in New Zealand. The effectiveness of asset-based policies is dependent

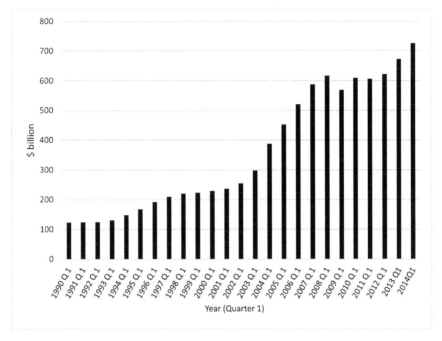

Figure 3.2 The value of the stock of private dwellings in New Zealand.
Source: retrieved on 11 May 2015 from www.rbnz.govt.nz/statistics/key_graphs/house_prices_values.

on the development of a financially sophisticated 'actuarial subject'. However, as homeowners become more sophisticated, they have developed strategies to resist policies that are aimed at getting households to pay for their welfare needs with their housing equity (Murphy 2012).

The rising value of the stock of housing in New Zealand is underpinned by significant house price inflation. In line with other economies that have experienced considerable mortgage market liberalisation, house prices have risen rapidly since the 1990s and there is considerable political and media discussion concerning the problems of housing affordability (Murphy 2014). The experience of rapid house price appreciation aligns well with an emerging narrative that positions all homeowners as beneficiaries of 'unearned' 'windfall gains' from the housing market. Under such conditions, it seems appropriate to consider homeownership as a suitable surrogate for a pension. However, this narrative does not take account of geographical variations in the performance of house prices.

Figure 3.3 charts real median house price changes nationally and across four cities over the period 1990 to 2013. This figure illustrates the operation of temporal and spatial processes that affect homeowners' potential capital gains. While all of the markets have increased over time, their rate of growth

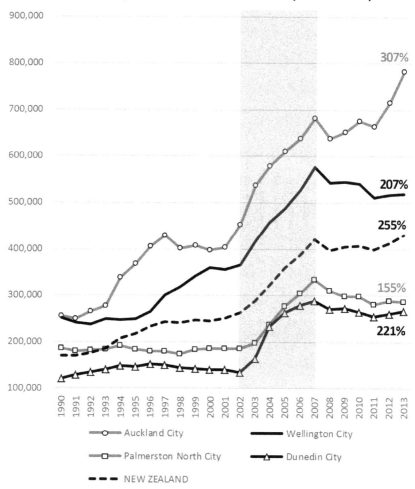

Figure 3.3 House prices in New Zealand.
Source: specially commissioned data set of stand-alone house prices from CoreLogic.

varies considerably (Auckland City 307 per cent, Palmerston North, 155 per cent). Prior to 2000, the four markets experienced divergent price trajectories, with Auckland and Wellington experiencing rising prices and Dunedin and Palmerston North prices being somewhat static. From 2002 to 2007 all four markets experienced rapid house price gains. Following the global financial crisis, three of the markets have remained below their 2007 peak, whereas Auckland City has experienced a significant surge in real prices. These variations in house price dynamics reflect the intersection of broad macroeconomic processes and local housing market conditions. Significantly, for the purpose of our argument, these data are illustrative of the significant variations in local house prices dynamics at work within a national market that has been characterised by continuous upward price movements. Moreover, our focus on four urban centres, while illustrative of the complex dynamics of price movements, masks significant intra-urban and neighbourhood-level house price processes. Within a political and media discourse that has focused on the inexorable rise of house prices, the existence of substantial geographical variation in house price dynamics points to the complexities of the lived realities of homeownership and represents another challenge to the narratives that support asset-based welfare policies.

In line with international trends, the intersection of contemporary demographic processes (an ageing society), welfare dynamics (debates over national superannuation and asset testing) and house price trends, clearly situates homeownership in New Zealand within a set of new ideological and political understandings, based on wealth accumulation/decumulation and notions of asset-based welfare. Within the context of rising house prices and high homeownership rates among older households, it would seem that homeowners in New Zealand are well placed to assume the role of competent 'actuarial subjects' responsible for their welfare needs in old age. However, on closer inspection, the transition to an asset-based welfare regime is less clear-cut. While asset testing is embedded in government policies around residential care subsidies, a significant number of homeowners are resisting the policy by placing their homes in family trusts. The traditional role of outright ownership, as a barrier against income poverty, continues to play an important role for older households in New Zealand and reverse mortgage products are not well developed. Significantly, in contrast to popular and political understandings of house price changes in New Zealand, that posits a uniform experience of capital appreciation, we have shown how spatio-temporal processes affect the experience of homeowners in different parts of the country.

Conclusions

In this chapter we have actively engaged with issues relating to the meaning and practices of homeownership and the wider dynamics of politics and economics (Cook *et al.* 2015). We have argued that, in contrast to the relative 'material fixity' of housing, popular and political understandings of home and homeownership, as well as the multiple experiences of owning, have been undergoing significant transformations. In this chapter we have focused our attention on the manner in which homeownership has been implicated in emerging debates around asset-based welfare. Where once homeownership was promoted in the United Kingdom, Australia and New Zealand within a discourse of 'savings', 'citizenship' and 'security', increasingly homeownership is embedded in narratives of wealth accumulation and decumulation. The emerging discourse of homeownership as a measure of 'personal achievement' and 'self-responsible neo-liberal citizenship' is increasingly aligned with a policy imaginary that positions homeowners as competent 'actuarial subjects', actively engaged in calculative practices designed to secure their future welfare. Underpinning this transformation in the ideology of homeownership are a set of empirical processes centred on rising house prices, increases in equity withdrawal and the increasing significance of housing wealth for households. Consequently, these evolving discourses are implicated in the material practices of homeowners and their rationales for entering homeownership and engaging in financial practices, such as equity withdrawal (Smith 2008, 2015).

Notwithstanding the increasing 'fungibility' of housing as an asset, we have argued that the logic of positioning homeownership as a central component of an asset-based welfare system is problematic and masks issues of socio-economic inequality within the tenure. The benefits of homeownership are unevenly distributed and housing markets have been shown to reinforce, rather than diminish, wealth inequalities (Malpass 2008; Smith 2015). To explore the complexities of emerging understandings of homeownership and asset-based welfare policy development, we have examined some key trends shaping homeownership in New Zealand. At one level, a set of demographic and housing trends clearly positions New Zealand within emerging discourses and practices centred on housing wealth accumulation/decumulation and asset-based welfare. However, the evolving experiences and practices of homeowners in New Zealand reveal a complex set of processes of resistance and inclusion/exclusion. Older homeowners are actively resisting asset-based welfare policies, but are also benefiting from housing wealth accumulation. Reflecting the geography of house prices, some homeowners are benefiting significantly (intentionally or accidentally) from rising house prices, while others are experiencing less dramatic wealth accumulation. Moreover, declining homeownership rates point to processes of exclusion from the sector that represent a long-term threat to the assumed benefits of an asset-based welfare system.

From our review of contemporary debates concerning the role of housing in emerging asset-based welfare systems, it is clear that the discourses and practices surrounding homeownership are being reformulated. Significantly, these discourses and practices are shaping government policy debates in Australia and New Zealand. In this chapter, we have argued that policy-makers are increasingly constructing homeowners as actuarial subjects engaged in seemingly unproblematic processes of wealth accumulation/ decumulation. However, in reality, homeownership is characterised by a multiplicity of 'wealth accumulation' outcomes (including losses). Moreover, our argument highlights the extent to which people are differently placed in relation to how well their home – as an assemblage of money and materials – works as a safety net or as a welfare substitute. We argue that the manner in which older homeowners are engaging with, or resisting, emerging asset-based welfare policy practices needs to be investigated in order to understand how contemporary homeownership is being reconstituted under neoliberalism.

Acknowledgements

This work was undertaken as a component of the Ministry of Business, Innovation and Employment public good science funded research program 'Finding the Best Fit', led by the Centre for Research, Evaluation and Social Assessment (CRESA) with assistance from the BRANZ Levy and the Commission for Financial Capability.

References

Aalbers, M.B. (ed.). 2012. *Subprime Cities: The Political Economy of Mortgage Markets*. John Wiley & Sons, Ltd, Chichester.

Castles, F.G. 1996. Needs-based strategies of social protection in Australia and New Zealand. In G. Esping-Andersen (ed.) *Welfare States in Transition: National Adaptations in Global Economies*. Sage Publications with United Nations Research Institute for Social Development, London.

Colic-Peisker, V., Ong, R. and Wood, G. 2015. Asset poverty, precarious housing and ontological security in older age: an Australian case study. *International Journal of Housing Policy* 15(2): 167–186.

Cook, N., Davison, A. and Crabtree, L. 2015. The politics of housing and home. In N. Cook, A. Davison and L. Crabtree (eds) *Housing and Home Unbound*. Routledge, London.

Doling, J. and Ronald, R. 2010 Property-based welfare and European homeowners: how would housing perform as a pension? *Journal of Housing and the Built Environment* 25(2): 227–241.

Forrest, R. and Hirayama, Y. 2015. The financialisation of the social project: embedded liberalism, neoliberalism and home ownership. *Urban Studies* 52(2): 233–244.

Forrest, R., Murie, A. and Williams, P. 1990. *Home Ownership: Differentiation and Fragmentation*. Unwin Hyman, London.

Fox O'Mahony, L. and Overton, L. 2015. Asset-based welfare, equity release and the meaning of the owned home. *Housing Studies* 30(3): 392–412.

Gotham, K. 2009. Creating liquidity out of spatial fixity: the secondary circuit of capital and the subprime mortgage crisis. *International Journal of Urban and Regional Research* 33(2): 355–371.

Groves, R., Murie, A. and Watson, C. (eds). 2007. *Housing and the New Welfare State: Perspectives from East Asia and Europe*. Ashgate, Aldershot.

Immergluck, D. 2011. From risk-limited to risk-loving mortgage markets: origins of the U.S. subprime crisis and prospects for reform. *Journal of Housing and the Built Environment* 26 (3): 245–262.

Judd, B., Liu, E., Easthope, H., Davy, L. and Bridge, C. 2014. *Downsizing Among Older Australians*, AHURI Final Report No. 214. Australian Housing and Urban Research Institute, Melbourne.

Lovell, H. and Smith, S.J. 2010. Agencement in housing markets: the case of the UK construction industry. *Geoforum* 41(3): 457–468.

Malpass, P. 2008. Housing and the new welfare state: wobbly pillar or cornerstone? *Housing Studies* 23(1): 1–19.

Murphy, L. 2011. The Global Financial Crisis and the Australian and New Zealand housing markets. *Journal of Housing and the Built Environment* 26(3): 335–351.

Murphy, L. 2012. Asset-based welfare. In S.J. Smith, M. Elsinga, L. Fox O'Mahony, S. Ong, S. Wachter and C. Hamnett (eds) *International Encyclopaedia of Housing and Home, Volume 1*. Elsevier, Oxford.

Murphy, L. 2014. 'Houston we've got a problem': the political construction of a housing affordability metric in New Zealand. *Housing Studies* 29(7): 893–909.

New Zealand Treasury 2013. *The Future Costs of Retirement Income Policy, and Ways of Addressing Them*, draft paper for the Long-Term Fiscal External Panel, retrieved on 5 April 2015 from www.treasury.govt.nz/government/longterm/externalpanel/pdfs/ltfep-s3-01.pdf.

Ong, R., Haffner, M., Wood, G., Jefferson, T. and Austen, S. 2013a. *Assets, Debt and the Drawdown of Housing Equity by an Ageing Population*, AHURI Positioning Paper No. 153. Australian Housing and Urban Research Institute, Melbourne.

Ong, R., Jefferson, T., Wood, G., Haffner, M. and Austen, S. 2013b. *Housing Equity Withdrawal: Uses, Risks, and Barriers to Alternative Mechanisms in Later Life*, AHURI Final Report No. 217. Australian Housing and Urban Research Institute, Melbourne.

Parkinson, S., Searle, B.A., Smith, S.J. Stoakes, A. and Wood, G. 2009. Mortgage equity withdrawal in Australia and Britain: towards a wealth-fare state? *European Journal of Housing Policy* 9(4): 365–389.

Price, D. and Livsey, L. 2013. Financing later life: pensions, care, housing equity and the politics of old age. In G. Ramia, K. Farnsworth and Z. Irving (eds), *Social Policy Review 25: Analysis and Debates in Social Policy*. Policy Press, Bristol.

Ronald, R. 2008. *The Ideology of Home Ownership: Homeowner Societies and the Role of Housing*. Palgrave Macmillan, Basingstoke.

Saunders, P. 1990. *A Nation of Home Owners*. Unwin Hyman: London.

Saville-Smith, K. 2013. *Housing Assets: A Paper for the 2013 Review of Retirement Income*. Prepared for the Commission for Financial Literacy and Retirement Income. CRESA, Wellington.

Searle, B. and McCollum, D. 2014. Property-based welfare and the search for generational equality. *International Journal of Housing Policy* 14(4): 325–343.

Searle, B. and Smith, S.J. 2010. Housing wealth as insurance: insights from the UK. In S.J. Smith and B.A. Searle (eds) *The Blackwell Companion to the Economics of Housing: The Housing Wealth of Nations*. Wiley-Blackwell, Chichester.

Sherraden, M. 2003. Assets and the social investment state. In W. Paxton (ed.) *Equal Shares: Building a Progressive and Coherent Asset-Based Welfare Policy*. IPPR, London.

Smith, S.J. 2008. Owner-occupation: at home with a hybrid of money and materials. *Environment and Planning A* 40(3): 520–535.

Smith, S.J. 2015. Owner occupation: at home in a spatial, financial paradox. *International Journal of Housing Policy* 15(1): 61–83.

Smith S.J. and Searle B.A. (eds). 2010. *The Blackwell Companion to the Economics of Housing: The Housing Wealth of Nations*. Wiley-Blackwell, Chichester.

Smith, S.J., Searle B.A. and Cook, N. 2008. Rethinking the risks of homeownership. *Journal of Social Policy* 38(1): 83–102.

St John, S., Dale, M.C. and Ashton, T. 2012. A new approach to funding the costs of New Zealand's ageing population. *New Zealand Population Review* 38: 55–76.

Statistics New Zealand 2014. *2013 Census QuickStats About Housing*. Retrieved from www.stats.govt.nz.

4   Building on sand?

Liquid housing wealth in an era of financialisation

Fiona Allon and Jean Parker

In November 2014, Phil Chronican, CEO of ANZ – one of Australia's four leading banks – described Australians as having 'a bit of an irrational obsession with housing as an investment class' (Yeates 2014). With 60 per cent of Australia's household wealth embodied in housing (the second highest rate in the world after Norway) and housing debt at historic highs of $1.84 trillion (AUD) or the equivalent of $79,000 per person (ABS 2014), Chronican's comments propose an explanation of a real shift. However, the increasing centrality of real estate investment in Australian society cannot be adequately understood at the level of personal obsession or individual irrationality (Allon 2008). Nor can it really be understood as the result of an episode of 'irrational exuberance',[1] the phrase used to describe the speculative stock market bubble of the 1990s and the psychological contagion that caused the excessive investor enthusiasm so characteristic of the time (Shiller 2005, 1–2).

When we look at the suite of economic and housing policy changes that have encouraged the investment behaviour of Australian home-buyers since the 1980s, we find nothing irrational in their focus on residential real estate. Rather, their choices demonstrate a sound understanding of the kind of investments necessary in order to plan for an uncertain future, and reflect the new roles that housing wealth, in particular, now plays in their lives. No longer is homeownership simply a matter of shelter in one's working life or even into retirement. Instead, it has become an important strategy of asset acquisition for people who increasingly view themselves as financial subjects taking responsibility for their own financial futures, primarily by investing in property assets such as housing that will hopefully appreciate in value and provide a fungible source of wealth (Smith 2008). In turn, such assets have been redefined as an essential form of self-provisioning, indeed as an important guarantee of life security for many at a time when the state has withdrawn its commitment to guarantee various kinds of social provision, be it housing or education or health. Alongside these shifts, housing has acquired new registers of emotional attachment for ordinary individuals and households, grounded in the imperative to accumulate assets as a means of individualised risk management. This represents an emergent form of housing subjectivity, one characterised by the incorporation of financial calculation at the level of identity and self-actualisation. The recent redefinition of housing not only as an 'asset-class' but also as liquid housing wealth that can be accessed at any time, therefore provides a site on which new and emerging forms of capitalist social relations, distinct from the so-called investor euphoria or speculative manias of the past, can be observed.

This chapter examines the transformation of housing into a site of 'liquidity' whereby the capital tied up in housing and in households more generally is 'unlocked' as a liquid financial exposure. In doing so it contributes to the project of unbounding housing/home by exploring the financial mechanisms that underpin the increasing fungibility of housing wealth. It addresses the consequences of this shift for ordinary homeowners, for government as well as for financial markets. In particular, it focuses on the implications of *in-situ* home equity withdrawal through two lenses. The first lens focuses on government and other state institutions. It draws attention to ways in which government policies are being shaped in response to the potentialities of home equity withdrawal. We will examine two developments that are indicative of the shift: the role of home equity extraction in boosting overall macroeconomic performance, and also the drive to mobilise housing wealth as a substitute for government-funded welfare services.

First, we argue that home equity withdrawal, facilitated by 'flexible' mortgage products, has been playing an important role in unlocking a stream of household spending that has been treated as a form of economic stimulus by governments. We look at the significant macroeconomic role of this stimulus, and suggest that this has encouraged governments and central banks to nurture housing investment through policy. Second, we examine the ways in which home equity withdrawal is increasingly being looked to by governments and individuals as a privatised source of income. The chapter argues that home equity is one key way in which neoliberal governments are seeking to shift the costs and the risks previously borne by the welfare state on to individuals.

In interaction with economic and policy changes around housing, then, we are seeing homeowners treating their mortgage as an all-purpose source of spending and security. However, this development is far from simple or straightforward. Rather than a singular or univocal process that magically releases equity for consumption (as it is often portrayed), equity withdrawal must be recognised as an unpredictable, paradoxical space in which benefits are contingent, risks are variable, and in which the desired aim of security is in fact frequently undercut by increased indebtedness and financial fragility. In exploring these contradictions, we follow the lead of Aalbers and Christophers (2014, 2–3) who argue that current research needs to consider the way in which 'housing is implicated in the contemporary political economy in numerous, critical, connected, and very often contradictory ways'. Our aim is to contribute to this project, and to explore the many contradictions that arise between the *liquidity* enabled by financial innovation and the fundamental *illiquidity* of housing as a social good.

Liquid assets?

> housing wealth, which has traditionally been regarded as a fixed asset, is increasingly fungible. It is no longer trapped behind bricks and mortar: it can easily be rolled out – literally dematerialised – and spent on other things.
>
> (Smith and Searle 2008, 39)

As a result of financial innovations in the 1980s and 1990s, mortgage holders found themselves able to fund consumption and investment out of the equity embodied in their homes. These innovations and their impacts on the global economy have been discussed extensively in the aftermath of the subprime mortgage collapse (Allon and Redden 2012; Brenner 2009; Finlayson 2009; Greenspan and Kennedy 2008; Immergluck 2011; Lapavitsas 2014; Lowe 2011; Yates 2014; Yates and Berry 2011). For our purposes there are two aspects of this picture that are key. The first is the global house price boom. As Lowe *et al.* (2012, 107) point out, one of the main consequences of the combination of cheap credit and access to global capital after financial deregulation was a 'surge in house prices almost everywhere lasting for nearly three decades, rising sharply after 1995 and peaking between 2000–2005'. Between 2000 and 2005 the value of residential property in the developed economies rose by the equivalent of 100 per cent of those countries' GDPs, a dramatic appreciation of house prices which *The Economist* later dubbed 'the biggest bubble in history' (Brenner 2009, 52).

The global price boom is also a good indicator of the extent to which the circulation of capital within the financial system more broadly had become increasingly grounded in housing, and by extension, in the ordinary business of households taking out mortgages and buying and selling homes. After all, it was such home loans that provided the collateral for the mortgage-backed securities (MBS) that had become a substantial source of liquidity within financial markets, as well as the preferred form of long-term investment in the portfolios of hedge funds, mutual funds and pension funds around the world. Attracting a higher rate of interest than Treasury Bills, MBS were in great demand by international investors in the constant *search for yield*. Escalating investor demand for securitised mortgages therefore played a direct role in driving up housing prices across the globe. On the other side of this profitable mortgage debt regime was the huge expansion of householders for whom investing in housing, and home ownership in particular, was made possible for the first time. Millions of homeowners across the globe were brought into a new kind of spatial and temporal proximity with the financial markets that were thriving on repackaging and reselling their mortgages. And in turn, more householders were targeted for loans as the result of the seemingly insatiable appetite of investors for MBS. Coinciding with the dramatic expansion of mortgage markets and the availability of new, ever more diverse kinds of mortgage products, including subprime and adjustable-rate mortgages, the trade in securitised mortgage debt was readily satisfied by the swelling ranks of homeowners who also hoped to 'bank' on housing for their future financial security (Lowe *et al.* 2012).

At the same time, innovative mortgage products such as home equity lines of credit, redraft facilities and reverse mortgages allowed households much greater control over the equity embodied in their homes (Cook *et al.* 2009). For the first time the home was no longer simply a long-term investment, or a spatially fixed, and largely inert, illiquid capital asset – a point to which we will return later in the chapter. Instead, these products allowed homeowners from across the income spectrum to tap into the rising equity in their homes in order to fund consumption or other welfare needs, or as a safety net for unexpected emergencies. As a result, households were drawn into financial markets both through their conversion of equity into increased mortgage debt and their use of new financial instruments. Their daily life was significantly transformed by the "global mortgage market with thousands of new products that were the conduit for the connection of household budgets to a tsunami of footloose capital" (Lowe *et al.* 2012, 113).

If financial markets shaped the global dynamics of the housing price boom, they also coloured the way in which home equity increases (through rising house prices) interacted with domestic economies. In fact, it is now beyond question that home equity extraction, made possible by novel financial products, played a key role in transmitting the housing boom into the broader economy and thus facilitating wider economic growth. In a period of stagnant wage growth in key economies (notably the United States, United Kingdom and much of Europe), the option of drawing into home equity without selling, and before having paid off mortgage debt, became not only possible but increasingly important for both household and national economies. Indeed, a number of political economists have argued that consumption fuelled by the extraction of home equity through new mortgage products was at the heart of economic growth in the developed world during the boom years from 2000 to 2007. Richard Brenner, for example, calculates that 98 per cent of the US GDP increase between 2001 and 2006 consisted of personal consumption and residential investment (Brenner 2009, 40). If, as the data suggest, much personal consumption during this time was boosted by credit released via housing equity withdrawal, the role of housing in the pre-global financial crisis (GFC) global economy was central. In the United Kingdom, similarly, and as Lowe *et al.* (2012, 111) note, at its peak (between 2003 and 2006) housing equity withdrawal funded 9 per cent of consumer spending (see also Froud *et al.* 2011). In the context of wage stagnation in both countries, then, this source of consumption funds was particularly significant. Moreover, Australian research has demonstrated that a one-dollar rise in housing wealth leads to a three-cent increase in non-housing consumption (see Windsor *et al.* 2013), a trend that was amplified in the United States, where a 7–8 cent rise in consumption follows every dollar increase in house value (Brenner 2009, 40). During the pre-crisis housing market upswing, therefore, home equity withdrawal provided an astonishing amount of economic stimulus to the global economy, confirming the extent to which housing, perhaps surprisingly, 'turns out to be of enormous significance for understanding capital circulation in the contemporary world' (Aalbers and Christopher 2014, 4).

Economists have long established that the 'wealth-effect' from rising house prices exceeds that of other assets (see Greenspan and Kennedy 2008, 4). In the early 2000s, however, the boost to consumption and economic growth was on a scale that seemed to go beyond the traditional housing wealth effect. A team of Australian Reserve Bank economists, for example, have proposed that on top of the traditional housing wealth-effects, home equity borrowing increased consumption by way of the provision of a new stream of credit to those households with access to mortgages (Windsor *et al.* 2013, 2–3). The amorphous impacts of the 'wealth effect' were therefore translated directly into tangible housing wealth, which had become important due to its vital role 'as collateral against which people can borrow to finance consumption' (Windsor *et al.* 2013, 2–3). Windsor and his team conclude that the aggregate increases in consumption which track the rise in house prices are directly attributable to the capacity of home equity loans to 'release' liquid funds through mortgage-backed debt. The real link, then, between housing price rises and increased consumption during this period was that:

> increases in home prices loosen credit constraints and therefore raise spending through an increase in the value of collateral, the opportunity for home equity redraws and/or through a reduction in the necessary level of buffer-stock, or precautionary, saving.
>
> (Windsor *et al.* 2013, 3)

It is more than likely that the exceptional role of home equity loans in Australian consumption pointed to by Windsor *et al.* is indicative of the global trend. Again we see the double impact of financialisation on boosting (for a time) macroeconomic performance: a housing boom driven by the demand for MBS, and a store of (fictitious) wealth created for householders as their equity was 'unlocked' and unleashed into domestic economies with the aid of new-generation home equity loans. However, Windsor *et al.* also allude to one of the main paradoxes that the financialisation of home brings into play. At the same time as there is an ongoing expectation of home as an ontological anchor on the one hand, there is also its emerging role as a class of financial asset on the other – and an especially risky one given the volatility of housing markets.

This tension is perhaps even more pronounced because both dimensions are bound up in the same material object, creating an inherent propensity for destabilisation at the centre of the presumed ontological role of home as security and stability. This tension is exacerbated when the fears that encourage people to look to housing investments for security go hand in hand, as Windsor *et al.* state, with the lowering of the personal savings rate. In these instances, the appreciation of assets creates an illusion of wealth and security that is in fact premised on greater levels of indebtedness and financial precarity. In other words, and as Johnna Montgomerie and Mirjam Büdenbender (2015, 392) make clear, 'residential housing can be a wealth-generating asset or a highly leveraged vehicle capable of decimating the household's entire financial security'. Moreover, with growing employment insecurity and the continuing precarisation of labour, the accumulation of assets is now promoted even when an intensified precarity works to install a paradoxical circularity whereby it is increasingly difficult to sustain any form of financial obligation, let alone the ongoing negotiation of increased indebtedness. In this sense, uncertainty and risk are thus laid out as the cornerstones of a secure future.

It is, however, the fundamental discrepancy between the liquidity of housing as a tradeable financial asset and its role as a relatively illiquid site of attachment and belonging that is the source of both profits *and* problems on financial markets. As Dick Bryan and Michael Rafferty (2014, 5) put it, 'There is a clear tension between, on the one hand, the inherent illiquidity of houses and households and, on the other hand, the liquidity demanded by financial markets'. In fact, it was precisely this tension (the bundling of mortgages into packages embroiled in global markets vs cultural expectations and the immediate lived experience of 'home') which became irreconcilable during the global financial crisis, triggering major faultlines in the residential capitalism that was expected to deliver ever-increasing opportunities for capital accumulation.

After all, it was precisely this expectation that structures economic policy at both macro and micro levels. In a research paper written for the Reserve Bank of Australia, for example, the year before the bursting of the sub-prime bubble, Luci Ellis examined the housing boom in Australia and globally. Her conclusion that financial innovation had created a magic-pudding of endless growth in housing wealth that would spur general consumption across the economy while simultaneously mitigating risk was not unique and in fact mirrored those of other central bankers at the time. In effect, this paradigm suggested that securitisation had changed the rules of financial risk. Or, as a heading in Ellis' paper put it, 'Old Rules of Thumb for Balance Sheets Might be Misleading'. As Ellis (2006, 29) argued: 'The most important lesson to draw from recent international experience is that a run-up in housing prices and debt need not be dangerous for the macroeconomy, was probably inevitable, and might even be desirable.'

Ellis's paper provides a useful insight into how those with their hands on the levers of the Australian economy understood the role of the growing housing sector for the performance of the whole economy. Specifically, she identifies the institutional features introduced by successive Australian governments that have facilitated the rise of the housing market. These include financial deregulation (which opened Australia to non-bank lenders that played a key role in lowering the costs of mortgage products in the 1990s) and a suite of tax measures (mortgage deductibility, capital gains tax exemptions and negative gearing). It is this set of economic and policy changes which has nurtured the investment behaviour of Australian home-buyers since the 1980s. The other aspect of state activity that is crucial in understanding growth of the Australian housing market is the interest-rate decisions made by the RBA itself. The lowering of the cash rate from 4.75 per cent in October 2011 down to record lows of 2 per cent in May 2015 spurred activity in the housing market that even the RBA Governor Glenn Stevens himself called 'crazy'.

Following the comments from Stevens and other Reserve Bank governors, public debate ensued over housing affordability and the prospects of a housing price bubble that might suddenly burst. Such debates make clear how seriously the major political parties take the protection of housing asset rises, and in turn the centrality of housing and households to financial stability more broadly. There are two aspects to these debates that are worth flagging here. First, there is a general fear that such concerns might undermine investor confidence and lead to a price crash. Second, they are acutely tuned to the apprehension of homeowners themselves about the prospects of their assets devaluing. Both elements can be seen in the 2015 episode. Even when serious concerns were raised about housing affordability, both major political parties still refused to countenance the view that prices are unsustainable or problematic. Indeed, in this instance, the leaders of both major parties seized the opportunity to claim responsibility for rising house prices. The political sensitivity of housing asset wealth can be clearly seen in comments from Australia's then Coalition Prime Minister Tony Abbott, who attacked Bill Shorten, the leader of the Opposition Labor Party, describing him as:

> a menace to the economic welfare of the people of Australia. Australians have mortgages and the last thing they want to see is the decline in their most important asset ... the Leader of the Opposition is saying ... people's houses are worth too much.... I certainly don't think we should be aiming for lower house prices, which seems to be the policy of the Labor Party, because lower house prices would mean a weaker economy.... Do not trust this man [the leader of the Labor Opposition] with your house price, do not trust this man with your superannuation, do not trust this man with your future.... Because what he wants is your house to be worth less.
>
> (Abbott, quoted in Borrollo 2015)

As well as the rhetorical support for continuous housing asset price growth, the commitment of the major parties to continued nurturing of the housing market can be seen in the ongoing support for the policy settings that encourage investment in housing. In the case of the 2015 debates this can be seen in the commitment to maintaining negative gearing tax incentives. In recent years, investors have accounted for a higher proportion of new loans in the housing market than owner-occupiers (Kohler 2015). Many commentators and economists argued that this was a disturbing development, both in terms of the stability of the housing market, and also in terms of its impact on housing affordability for first-time homeowners. However, despite these concerns, and even in the face of a strong need for budgetary savings and fiscal consolidation, negative gearing, along with other policies which directly favour housing investment price appreciation, was deemed to be too politically sensitive and remained untouched.

As these debates continued it became increasingly clear that the key macro-economic role being played by property investment in Australia and elsewhere has been accompanied by profound changes in how housing itself is now understood. In other words, homes have taken on new layers of highly contradictory meanings. Historically, housing's 'bricks-and-mortar' qualities have signified a robust and essentially stable form of physical, and indeed social, protection from the vicissitudes of a capitalist market economy. Indeed, it is the relative fixity of housing that actually enables it to provide its 'use value', that is as places to *live in*. However, onto these existing foundational meanings of housing and homeownership, the proliferation of new mortgage products has added novel and quite paradoxical layers of association (see Smith 2015). If the privileged role of housing was its ability to provide fixity, both as shelter and as a stable long-term investment, then it is now liquidity, the ability to dematerialise the home into a series of liquid financial exposures, that are then able to be converted into a source of money, that define its current position. Home-equity loans, interest-only loans, reverse mortgages and lines of credit have allowed mortgagees to 'liquidify' the value embodied in their houses in order to provide funds that enable them to consume today. In fact, in the context of steadily rising house prices and low wage growth, a section of homeowners have come to rely on this non-wage income stream.

If, for Marx, the key paradox of the commodity was that it was simultaneously a use value and an exchange value, concrete and abstract, particular and equivalent, in a manner that was never fully resolved or resolvable, then the financialisation of housing leads to an amplification of such paradoxical attributes. The use-value of the house continues to inform the financial commonsense that sees property ownership as a safe investment that provides a housing service over the life-course. The inexorable rise of the Australian housing market, even after the GFC, has done this commonsense no harm. However, the growing practice of equity withdrawal entails the prospect of owners retiring in debt in a way previously impossible. In this sense, the temporality of housing wealth has become completely upended and riddled with contradiction. Historically, housing wealth was something built up throughout a working life of paying wages into a mortgage. It was a 'nest-egg' for the future that would provide shelter during retirement. It would also provide a start to the asset-base of the next generation through their inheritance of the accumulated equity. The proliferation of mortgage-related financial products, however, has fundamentally changed both temporal and spatial patterns of home equity: instead of growing over the course of the mortgage, home equity is now 'leaking' (Parkinson *et al.* 2009) during precisely the period of the life-cycle where it traditionally accrued.

If, traditionally, housing wealth has been seen as an asset that could be drawn on reluctantly and as a last resort, now 'home equity is used as a *financial resource* and built up or released over the life course via *financial products* ... leaving little of the housing asset left when it is needed later in life' (Doling and Ronald 2010; our emphasis).

As evidence of this shift, a study by Parkinson *et al.* (2009, 379–380) found that:

> taking out the very youngest cohort of home buyers, the inclination to engage in equity borrowing increases with youth not age ... mortgage borrowing is bringing spend from housing wealth forward, not to the retirement or pre-retirement years, but rather to fund spending needs much earlier in the life-cycle.

Despite this trend, itself responsible for the historic diminishing of personal savings rates in economies like the United States and Australia, housing has not lost its earlier meanings of stability and security, or its perception as a secure investment for the future. Housing has instead become over-signified, with the traditional connotations of homeownership co-existing with new, and highly contradictory, qualities. Homeownership now has come to represent, simultaneously, a means of shelter for today, a 'credit-card' that can fund consumption, a form of insurance for loss of earnings, an asset for future income streams, the basis for other asset purchases, a bequest to future generations and a protection in old age. Of course, housing cannot, in fact, perform all these functions. But the experience of the housing price run-ups has conditioned us to act as if it can.

Given the importance of home equity consumption for the performance of the global economy, it is not surprising that academic research has begun to explore reasons for its extraction, the kinds of households that have relied on it to fund expenses (Smith *et al.* 2009; Parkinson *et al.* 2009; Wood *et al.* 2013), and the impacts of mortgage-led consumption on home (Cook *et al.* 2013). Still, there is more to learn about the motivations of those who have been able to liquefy and withdraw, ATM-like, part of the value of their homes (Klyuev and Mills 2006; Smith and Searle 2008, 2010; Smith *et al.* 2009). Some of the existing data indicate the importance of recognising class distinctions within home equity extracting households. Just as Windsor *et al.* (2013) stress the differential impact of house price rises on different age groups and those who are looking to buy and to sell, we must also note that the use of home-equity derived credit has different purposes for high and low income mortgagors. Housing investment may have flourished within the middle and working classes, but it appears there are key class distinctions in how such investment is used. Alan Greenspan and James Kennedy, for example, discuss these distinctions in reference to the United States. They argue that those that are 'liquidity restrained' (read lower-income earners) consume the wealth they withdraw from their homes, whereas the well-off use home equity withdrawal tools to free up funds for further investments:

> liquidity constrained households use the equity from cash out refinancings to fund current consumption, which results in a decline in their overall wealth. By contrast, home equity extracted by non-liquidity constrained households would be invested in other types of assets, resulting in no change in wealth ... liquidity constrained households converted two-thirds of every dollar of home equity removed in refinancing to consumption; non-liquidity constrained households did not use any of those funds for consumption.
>
> (Greenspan and Kennedy 2008, 4)

One stark face of this differentiation can be seen in the fact that 'a considerable portion of the equity extracted through cash out refinancings and home equity loans was used to repay non-mortgage debt, largely credit card loans' (Greenspan and Kennedy 2008, 2). This use of home equity loans points to a significant section of homeowners struggling to manage multiple debts. In fact, paying off credit card debt represented 27 per cent of the funds extracted with home equity loans in the United States, where $50 billion of consumer debt was paid with the use of withdrawn home equity between 1991 and 2005 (Greenspan and Kennedy 2008, 9).

We have seen the ways in which the extraction of home equity made possible by new mortgage products has been a source of consumption, and therefore of economic growth. We have also seen the ways in which Australian governments and the RBA have consciously cultivated housing wealth with some understanding of its centrality in economic performance. We now turn to discuss another possibility created by the capacity to withdraw home equity *in situ*, namely the provision of a private stream of non-waged income.

From welfare state to wealth-fare state: the role of housing

> The preponderance of housing points to a key factor behind the financialisation of households in recent decades: rising household indebtedness has been associated with changes in the social provision of basic services including housing, health, education, transport and so on.

To the degree to which social provision has retreated, or failed to expand, private provision has taken its place, mediated by finance.

(Lapavitsas 2014, 240)

We have seen how at the turn of the century the seemingly endless influx of cheap credit into housing markets led to a dramatic global surge in housing wealth. This housing wealth came to be seen by governments as a possible solution to pressures on the welfare state. This was the result of a convergence of factors, and stemmed primarily from the fact that from the mid-1970s most developed nations saw the creation of deep pockets of structural unemployment and disadvantage that had not existed for some decades. At the same time, ageing populations added extra demands on pensions and healthcare systems. In the emerging neoliberal framework, dealing with these challenges by increasing public provision and taxation was deemed to be unviable.

The solution, it was generally accepted, should not come from the expansion of the welfare state, but rather from the expansion of personal finance. Indeed, all of these factors converged in the notion that personal asset wealth, particularly housing, could be a way to fund services that had previously been provided publicly as part of the welfare state. Housing wealth came to be seen as a resource that could be tapped into in order to protect people's living standards in retirement, to fund the provision of (increasingly privately delivered) health and education services and even to sustain income through periods of unemployment. Specifically, the financialisation of housing facilitated a transformation whereby housing wealth was understood as the basis for a privately provided alternative to 'social security'. In particular, the capacity for mortgagors to withdraw housing equity *in-situ* (without selling) fostered a material shift that has allowed governments to propose a 'wealth-fare state' in the place of the welfare state (Lowe *et al.* 2012, 112; see also Doling and Roland 2010).

There are two dimensions to this development that are worth noting here. On the one hand, the cultivation of housing wealth as a form of welfare provision can be seen at the level of neoliberal government policy and philosophy. On the other hand, it is important to also discuss the changes in popular attitudes to housing that came to see investment decisions as providing a whole-of-life security blanket. The very interesting question also arises as to which of these shifts is driving the other. Castles has questioned whether the rise in homeownership has in fact facilitated the retrenchment of functions of the welfare state, or whether it followed those changes (Lowe *et al.* 2012). Unravelling this 'chicken and egg' (Lowe 2011, 207) conundrum is far from straightforward.

Either way, however, it is increasingly obvious that governments see in the housing wealth created by the boom, a privately supplied, privately funded alternative to the traditional welfare state. At the same time as cultivating private homeownership and withdrawing public alternatives, they are relying on the reality of ownership for owners themselves, to suggest a previously unthinkable retraction of government responsibility for welfare (Lowe *et al.* 2012, 113). In Australia, which is the primary focus of this chapter, persistent calls for reverse mortgages to substitute publicly funded pensions illustrate one way in which the financialisation of housing wealth is providing the scope for the retrenchment of the welfare state. As pension payments have become less adequate, particularly with the rise of private residential care, policy-makers have looked for ways to mobilise the wealth embodied in the housing assets of retirees. The long history of high homeownership in Australia combined with relatively low pension levels has created a particularly stark phenomenon of aged people who are asset-rich but liquidity poor. While not current policy of either major political party, there have been persistent calls for policy settings that would allow government to 'unlock' the estimated $625 billion (AUD) (Cowan and Taylor 2015) embodied in the housing stock of Australian retirees. Leading economic institutions including the Productivity Commission in 2012, the Commission of Audit in 2014 and the interim report of the Murray Financial System Inquiry have advocated reverse mortgages as a means of achieving this. As former NSW Labor Premier Kristina Keneally put it:

> Why are Australians so fearful of using the equity in the family home to fund retirement? I asked this question when I was NSW minister for ageing. 'You can't eat bricks and mortar' stakeholders told me. Technically that's true, but you can unlock the equity in a home in order to pay for food or whatever else an older person needs in order to live more comfortably in their retirement years.... Australia needs to face squarely the question of whether we can continue to ignore a lazy $600bn of equity just because we have an emotional obsession with the family home.
>
> (Keneally 2015)

In response to the recommendations of a 2012 Productivity Commission report, the Labor government developed policy to facilitate reverse mortgages to play a greater role in funding aged care provision (Johnson *et al.* 2013). However, as of 2015 only 8 per cent of retirees had a reverse mortgage (Cowan and Taylor 2015). Similarly, in the lead-up to the 2015 Federal Budget, the market-libertarian think-tank, the Centre for Independent Studies (CIS), produced modelling to show that by including the family home in the pension asset test, the government could move 70 per cent of pensioners off full payments. This, the CIS argued, would reduce the federal pension bill by $14.5 billion (AUD) annually (Cowan and Taylor 2015, 2).

At first sight the process of retirees funding their aged care through reverse mortgages (on housing wealth accumulated through their working lives) is perhaps not drastically divergent from longstanding notions of homeownership as a nest-egg built up through a working life. These mortgages fit within what Doling and Ronald describe as 'traditional' housing asset-based welfare (Doling and Ronald 2010). However, what is novel is the way in which the housing that forms the economic base of the proposed reverse mortgage programs is not envisaged *as housing*, but rather is seen as a fungible asset just like any other asset, and one that can be used as an income stream by retirees. This debate again reveals the fundamental ambiguity between the use-value of the house as a lifetime home and its financial value and purpose as an asset. Pensioners' houses may be the ones they've lived in for decades and raised families in, but with the aid of financial products like reverse mortgages, their homes can be 'liquified' and drip-fed back to them as an income stream even while they still reside in them.

The CIS report refers to this inevitable friction between the house as 'home' and as financial asset, noting that the family home 'as a retirement savings tool … invokes *complicated emotions*' (Cowan and Taylor 2015, 23; our emphasis). It is presumably these complicated emotions that have seen reverse mortgages remain a small part of the mortgage market. But once homes are reconceptualised as asset-streams, they become a potential private replacement for public welfare. This reconceptualisation is most recognisable when housing wealth is used interchangeably with superannuation wealth. In the lived reality of many pensioners the two are completely distinct and have different roles. However, thanks to the reverse mortgage, the lifetime home can become interchangeable with a financial asset, and once it is fungible, homes become indistinguishable from other sources of financial wealth. The equivalence is demonstrated by the CIS report, which is replete with statements such as:

> For most people, the two biggest assets they will own in their lives are their superannuation and their family home. It is no surprise that these assets form a significant core of savings people could use to support themselves in retirement.
>
> (Cowan and Taylor 2015, 21)

The focal points of the CIS intervention into the retirement incomes debate in Australia are particularly revealing. Cowan and Taylor identify two major impediments to the adoption of this solution to funding the incomes of Australian retirees. The first is the sentiment among many pensioners themselves that they are entitled to a government retirement income as a 'right', having paid taxes throughout their working lives. Unsurprisingly, given the overall aim is to transform the public pension system into a residual safety net, the CIS report rejects this belief, providing yet another clear sign that the politics of 'rights' that went along with the expansion of the welfare state is also inevitably one of the early victims of a policy shift towards self-provisioning.

One radical outcome of the CIS proposal would be that a generation of pensioners would substitute a public pension for 'eating the house', drawing on the equity in their homes to fund their retirement. Consequently, houses would no longer be passed down to relatives, but would become the property of the issuer of the reverse mortgage. This proposal raises very interesting questions about the traditional role of inheritance in homeownership societies, and suggests that the temporality of housing wealth is already undergoing significant transformation. We mentioned above that home equity withdrawal has been part of a greater trend to 'leakage' that sees a growing number of people retire with outstanding mortgage debt. The CIS plan would greatly accelerate the existing trend. Whereas historically mortgage debt was highest at the point of home purchase – early in the working lives of the owners – we are now contemplating a situation where (in the case of reverse mortgages) home equity levels peak towards retirement, after which debt increases, even continuing after death. This is the see-saw of equity extraction, a precarious balance between a wealth-generating asset for the lucky few, and years of crippling indebtedness and negative equity for many others. As such, and as Aalbers and Christopher (2014, 8) claim, it is not so surprising that now 'it is in housing that the vast wealth inequalities of capitalist societies … are often most visible and most material'.

Conclusion

A culture of financial calculation has developed around housing that explicitly encourages homeowners to participate as investors in the speculative appreciation of their homes. Consequently, a mortgage has become more than just a means of accessing homeownership and the long-term security it promises. It is now a way of acquiring an asset that can be leveraged in all kinds of ways: it can be 'flipped' in a rising market, provide a store of equity to be released for consumption, emergencies or to pay down outstanding debts, and serve as collateral for other loans. The home itself is increasingly viewed as a financial asset that can be strategically managed and traded. And in turn, this wealth now equates to much more than the steadily growing value of bricks and mortar; it is a capital base for hedging against the new risks and uncertainties of financialised futures.

The contradictory hopes and dreams that are now projected onto housing seem destined to surface in the future. The over-signification of housing as the 'solve-all' asset – for both government and homeowners – is so pregnant with contradictions that it seems inevitable that a reckoning must be on the horizon. In this sense, the ongoing public debates about housing affordability, market bubbles, interest rates and retirement incomes are only a prelude to the concerns yet to come.

Note

1 'Irrational exuberance' was the phrase used in 1996 by Chairman of the Federal Reserve Board, Alan Greenspan, to describe the speculative behaviour of stock market investors. See Shiller (2005).

References

Aalbers, M. and Christophers, B. 2014. Centring housing in political economy. *Housing, Theory and Society* 31(4): 373–394.

ABS. 2014. *4102.0 Australian Social Trends 2014*. Australian Bureau of Statistics.

Allon, F. 2008. *Renovation Nation: Our Obsession with Home*. University of NSW Press, Sydney.

Allon, F. and Redden, G. 2012. The global financial crisis and the culture of continual growth. *Journal of Cultural Economy* 5: 375–390.

Borrollo, E. 2015. Major political parties fire shots at one another over housing affordability. ABC News, 3 June. Retrieved on 22 October 2015 from www.abc.net.au/news/2015-06-03/major-political-parties-attack-over-housing-affordability/6516650.

Brenner, R. 2009. What is good for Goldman Sachs is good for America: the origins of the current crisis. Retrieved on 22 October 2015 from http://escholarship.org/uc/item/0sg0782h.

Bryan, D., Martin, R. and Rafferty, M. 2009. Financialization and Marx: giving labor and capital a financial makeover. *Review of Radical Political Economics* 41(4): 458–472.

Cook, N., Smith, S.J. and Searle, B. 2009. Mortgage markets and cultures of consumption. *Consumption, Markets and Culture* 12(2): 133–154.

Cook, N., Smith S.J. and Searle, B. 2013 Debted objects: homemaking in an era of mortgage-enabled consumption. *Housing Theory and Society* 30(3): 1–19.

Cowan, S. and Taylor, M. 2015. *The Age Old Problem of Old Age: Fixing the Pension*. Centre for Independent Studies. Retrieved on 22 October 2015 from www.cis.org.au/publications/research-reports/the-age-old-problem-of-old-age-fixing-the-pension.

Doling, J. and Ronald, R. 2010. Home ownership and asset-based welfare. *Journal of Housing and the Built Environment* 25: 165–173.

Ellis, L. 2006. *Housing and Housing Finance: The View From Australia and Beyond*. Research Discussion Paper, Reserve Bank Australia. Retrieved on 22 October 2015 from www.rba.gov.au/publications/rdp/2006/pdf/rdp2006-12.pdf.

Finlayson, A. 2009. Financialisation, financial literacy and asset-based welfare. *British Journal of Politics & International Relations* 11: 400–421.

Froud, J., Johal, S., Law, J., Leaver, A. and Williams, K. 2011. *Rebalancing the Economy (or Buyer's Remorse)*. Centre for Research on Socio-Cultural Change (CRESC) Working Papers Series (No. 87). Working Paper.

Greenspan, A. and Kennedy, J. 2008. Sources and uses of equity extracted from homes. *Oxford Review of Economic Policy* 24: 120–144.

Immergluck, D. 2011. *Foreclosed: High-risk Lending, Deregulation, and the Undermining of America's Mortgage Market*. Cornell University Press, Ithaca, NY.

Johnson, D., Worthington, A. and Brimble, M. 2013. Building the pillars of retirement income: what role for housing equity in easing financial stress in later retirement? *7th Australasian Housing Researchers' Conference* 6–8 February, Esplanade Hotel, Fremantle, Western Australia.

Johnson, D.P., Worthington, A. and Brimble, M.A. 2015. Building the pillars of retirement income: what role for housing equity in easing financial stress in later retirement? In *Australasian Housing Researchers' Conference*. Retrieved on 22 October 2015 from https://business.curtin.edu.au/local/docs/ahrc13/Building-the-pillars-of-retirement-income-what-role-for-housing-equity-in-easing-financial-stress-in-later-retirement.pdf.

Keneally, K. 2015. Morrison's pension reform is tinkering at the edges: there is another solution. *Guardian Online* (Australian edition) 7 May. Retrieved on 22 October 2015 from www.theguardian.com/commentisfree/2015/may/07/morrisons-aged-pension-reform-is-tinkering-at-the-edges-there-is-another-solution.

Klyuev, V. and Mills, P.S. 2006. *Is Housing Wealth an 'ATM'? The Relationship Between Household Wealth, Home Equity Withdrawal, and Saving Rates.* IMF Working Paper. Retrieved on 21 October 2015 from http://dx.doi.org/10.5089/9781451864229.001

Kohler, A. 2015. Personal website. Retrieved on 29 October 2015 from www.alankohler.com.au/?page=26

Lapavitsas, C. 2014. *Profiting Without Producing: How Finance Exploits Us All.* Verso Books, London.

Lowe, S. 2011. *The Housing Debate.* Policy Press, Bristol.

Lowe, S.G., Searle, B.A. and Smith, S.J. 2012. From housing wealth to mortgage debt: the emergence of Britain's asset-shaped welfare state. *Social Policy and Society* 11: 105–116.

Montgomerie, J. and Büdenbender, M. 2015. Round the houses: homeownership and failures of asset-based welfare in the United Kingdom. *New Political Economy* 20(3): 386–405.

Parkinson, S., Searle, B.A., Smith, S.J., Stoakes, A. and Wood, G. 2009. Mortgage equity withdrawal in Australia and Britain: towards a wealth-fare state? *European Journal of Housing Policy* 9: 365–389.

Shiller, R. 2005. *Irrational Exuberance.* 2nd edition. Princeton University Press, Princeton, NJ.

Smith, S.J. 2008. Owner occupation: living with a hybrid of money and materials. *Environment and Planning A* 40: 520–535.

Smith, S.J. 2015. Owner occupation: at home in a spatial, financial paradox. *International Journal of Housing Policy* 15: 61–83.

Smith, S.J. and Searle, B.A. 2008. Dematerialising money? Observations on the flow of wealth from housing to other things. *Housing Studies* 23: 21–43.

Smith, S.J. and Searle, B.A. 2010. *The Blackwell Companion to the Economics of Housing: The Housing Wealth of Nations.* John Wiley & Sons, Malden, MA.

Smith, S.J., Searle, B.A. and Cook, N. 2009. Rethinking the risks of home ownership. *Journal of Social Policy* 38: 83–102.

Windsor, C., Jaaskela, J. and Finaly, R. 2013. *Home Prices and Household Spending.* Discussion Paper, Reserve Bank of Australia. Retrieved on 22 October 2015 from www.rba.gov.au/publications/rdp/2013/pdf/rdp2013-04.pdf.

Wood, G., Parkinson, S., Searle, B. and Smith, S.J. 2013 Motivations for equity borrowing: a welfare-switching effect. *Urban Studies* 50 (3): 2588–2607.

Yates, J. 2014. Protecting housing and mortgage markets in times of crisis: a view from Australia. *Journal of Housing and the Built Environment* 29: 361–382.

Yates, J. and Berry, M. 2011. Housing and mortgage markets in turbulent times: is Australia different? *Housing Studies* 26: 1133–1156.

Yeates, C. 2014. Negative gearing blamed for irrational property obsession: ANZ boss Phil Chronican. *Sydney Morning Herald*, November 3. Retrieved on 22 October 2015 from www.smh.com.au/business/banking-and-finance/negative-gearing-blamed-for-irrational-property-obsession-anz-boss-phil-chronican-20141102-11fnax.html.

5 Cohabiting with cars

The tangled connections between car parking and housing markets

Elizabeth Jean Taylor

Car parking space: unnoticed but ubiquitous

Car parking occupies as much as 40 per cent of urban land in many cities, yet manages to go 'expected but unnoticed'. Ben-Joseph (2012, 135) observes how we 'demand convenient parking everywhere we go, and then learn not to see the vast, unsightly spaces that result'. Although valued for the sense of personal autonomy it offers, private car transport is ultimately dependent on there not being too many other cars around and on adequate car parking space. Given cars are stationary 95 per cent of the time (Vanderbilt 2008), the provision of convenient car storage is essential to car-based mobility (Hagman 2006; Pandhe and March 2012).

Notwithstanding spatial and other variations, trips by private car have been the predominant transport mode in Australian cities for several decades (Mees and Groenhart 2012). Over 90 per cent of households in, for example, the state of Victoria owned at least one car in 2011; and over half owned two or more cars (ABS 2011). For several decades the rate of increase in the number of cars has exceeded growth in population or households (ABS 2013). Reflecting this, Australian cities encompass extensive post-war car-oriented suburban developments, and have (not without conflict, see Davison and Yelland 2004) reconfigured older inner urban areas to accommodate the movement and parking of cars.

One of the urban places where provision is more or less compulsorily made for car storage is in the home. Materially, car parking has come to play an important role in the physical design of Australian dwellings. Double undercover garages, for example, are a standard feature integrated into new house designs (Mukhija and Shoup 2006). Car parking is further linked with housing through legal and regulatory mechanisms integrating car parking space with dwelling space. In an integration so ubiquitous as to be more or less invisible, purchasing a dwelling almost invariably entails purchasing infrastructure intended for – and in some cases legally restricted to – cars.

This chapter explores some of the ways in which housing and car parking are bound together, materially and through the production and sale of housing. It documents the extent to which car parking is a fundamental part of housing sales in an Australian city (Melbourne). A case study of two adjoining Melbourne housing developments built without on-site car parking is then given. Drawing on these cases, I argue that in order to facilitate the provision of housing without car parking, a series of overlapping linkages – social, economic, legal, political, architectural, historical and cultural – needed to be purposefully untangled and reconfigured. The effort required in doing so is indicative of the extent to which the default state of housing in Australian cities is bound up with car-based transportation.

Which comes first: parking space or the car?

Private cars have been entangled with urban development, and with the production of housing and 'home', for much of the twentieth century. Parking space is one manifestation of this entanglement, both a result and facilitator of high levels of car ownership and use.

An Australian Bureau of Statistics (ABS 2012) survey on waste management, transport and motor vehicle usage reports that 'not having a service available at all' was one of the main reasons why people did not use public transport in Australia (30 per cent); as well as lack of availability of a service at the right or convenient time (23 per cent). Studies, including from Australia, point to areas in which car use and associated parking provision is inelastic (Dodson and Sipe 2007; Motte-Baumvol *et al.* 2010; Zhao *et al.* 2013). For example, lower-income groups are sometimes restricted by housing market pressures to areas with poor transport alternatives. From this perspective, parking is essential to living in a car-oriented or car-dependent area.

Transport literature canvasses the possibility that rather than influencing travel choices directly, local transport infrastructure – including parking availability and price – influences where people choose to live (Bagley and Mokhtarian 2002; Cao and Cao 2014). Overlapping literature considers personal and cultural attitudes toward cars, including the comfort and privacy they may offer (Featherstone 2004; Gardner and Abraham 2008; Kent 2014). From this perspective, demand for automobility determines the spaces given to parking, irrespective of alternative transport provision.

Whether cause or effect, parking is critical to maintaining car-based transport – and is rarely openly priced. Often the only reason why people will not drive or own a car is a lack of free parking (Guo 2013; Guo and Ren 2013; Hagman 2006; McDonnell *et al.* 2010; Pandhe and March 2012; Shoup 2005). Recent critical interest has questioned the broader transport effects of parking policies that 'predict and provide' minimum parking that then appears to be 'free' despite its direct and broader costs (Shoup 2005; Litman 2006; Marsden 2006; Mingardo *et al.* 2015). Critics have suggested parking is oversupplied by not being openly priced (Guo 2013; Pierce and Shoup 2013; Shoup 2005; Wilson 2013); and, in the case of residential parking, that housing regulations that stipulate minimum parking provision reduce housing choice and affordability (Guo and Ren 2013; Li and Guo 2014; Manville 2013; Manville *et al.* 2013; McDonnell *et al.* 2010; Shoup 2005).

Tensions around parking space are experienced in intensifying Australian cities, as competition for space increases. Australian metropolitan plans have promoted the compact city as a means to achieve greater transport sustainability and housing affordability. However, they have also retained existing levels of, and policy approaches to, car parking. In cities undergoing spatial intensification, housing and parking density are increasingly in competition – physically and economically. Car parking policy is thus of relevance to many issues grappled with in the Australian urban literature, including: opposition to higher density housing and intensification; concerns about housing affordability; housing standards; and the sustainability of urban form (Bunker *et al.* 2002; Cook *et al.* 2013; Dodson and Sipe 2007; Easthope and Randolph 2009; Gurran 2008; Ruming and Houston 2013; Searle and Filion 2011; Woodcock *et al.* 2011). Even if transport patterns were to shift, cars and their parking spaces remain deeply embedded into the physical fabric of Australia housing, into ideas of home-making, and into regulatory frameworks for housing production.

Cohabiting: the car at home

One material form of bundling of car parking with housing is the garage and its integration into the design of newer, lower-density houses. In suburban detached housing, residential car spaces have moved over time to be undercover and eventually to be internally part of the house itself (Mouden 1992). Suburban development in Australia increasingly entails larger dwellings, smaller blocks and larger garages – at the expense, as lamented by Hall (2008), of gardens. Garage spaces account for a significant portion of modern house footprints and facades, and are one feature of the derided 'McMansion' or 'Garage Mahal' (Nasar *et al.* 2007).

The integration of housing and transport systems is partly determined by housing industry structures, which vary greatly by country (Ball 2003). Briefly, Australian housing suppliers of suburban, infill and high-rise housing operate in largely separate industries. The suburban house-building industry includes developers – who package land and infrastructure into estates, and sell land plots to homebuyers – and sub-contracted house builders (Dalton *et al.* 2013). Homebuyers select from housing designs from builders approved by the estate developer, with variance from standard designs attracting significant additional costs (Goodman *et al.* 2010). As at July 2015, based on listings on real estate website Domain.com.au, of 4,130 new houses and new house and land packages for sale in Melbourne, all had garages and most (3,561) had double garages or larger.

The garage as part of housing in part reflects location– the lack of alternatives to car-based transport in the urban fringe areas where new detached housing development predominates (Dodson and Sipe 2007). Garages also reflect social and consumer norms of what is to be housed. The storage of people and of cars are inextricably mixed in the marketing of both housing and cars. Housing is designed and marketed for ease of car access, and cars are marketed as homes – as, to quote one recent campaign, a 'great place to raise a family'. Housing developers in Melbourne also apply design guidelines of varying legal force – ranging from published guidelines to binding covenants on titles. At least one estate design guideline in Melbourne in 2015 (Manor Lakes 2015) stipulates that all new homes must include an enclosed garage. Beyond the requirement for a garage, the use of garage space in lower-density development is comparatively unregulated. Mostly garages store cars – however, the versatility of the large, utilitarian space is suggested by the term 'garage band' (Kauppila 2005) and by vaunted myth-making tales of technical entrepreneurialism in garages (Fuller 2015; Godelier 2007).

In medium-density and apartment housing the integration of car parking space is both material and legal. In multistorey apartment buildings, parking spaces are supplied in basement or lower levels, and are physically separate from dwellings. They may not be contiguous to dwelling spaces but are, however, usually 'bundled' onto the same property title as dwellings. Prior to the introduction of Strata Title in the late 1960s, facilitating ownership of single units in multi-unit developments (Randolph 2006), parking associated with 'flat' developments was not directly linked to dwelling titles. Nonetheless, during the 1960s, 'flat' development was designed around shared on-site car parking. Lewis (1999, 90–91) critically describes the six-pack as 'blocky boxes of two or three storeys on concrete stiles over a car park'.

Only older housing – largely meaning in Australian cities Victorian and Edwardian period terrace housing, and interwar 'flats' – does not physically integrate car parking. It is in these older dwellings that residents have residential use rights to on-street parking space. Such rights tend to be perceived as property ownership rights. In a recent Melbourne dispute, when threatened with a reduction of automatic on-street parking rights, a public response was to assert that 'you cannot take away people's rights to have a car' (*The Age*, 2 September 2015).

One mechanism for integrating parking and housing in new higher-density forms is the development approval process. Throughout the twentieth century, a variety of regulatory and social rules emerged to bring order to the spatial challenges accompanying increasing car ownership in cities. Designated parking spots, meters, time limits and off-street parking requirements all emerged as efforts to allocate and manage public parking resources and free up traffic flow (Marsden 2006; Marusek 2011). Traditional land use zoning approaches mandate minimum levels of parking provision for new developments, with minimum parking ratios entrenched in statutory planning. Australia is an exemplar of this parking policy situation. The stated purposes of the Victorian car parking policy include both 'to ensure the provision of an appropriate number of car parking spaces having regard to the demand likely to be generated' and 'to support sustainable transport alternatives to the motor car'.

Clause 52.06 of the Victoria Planning Provisions (VPPs), state-standardised planning scheme mechanisms, stipulates parking policy and ratios. In Victoria, for new developments, each new one-bedroom dwelling is to be provided with one off-street car parking space, and each two or three bedroom dwelling with two off-street spaces. For every five new dwellings, one visitor parking spot is to be provided. It is possible to reduce these parking ratios, but a permit for a discretionary 'waiver' is required. The applicant must satisfy the responsible authority of the 'appropriateness of allowing fewer spaces to be provided'. Higher-density – Strata Titled – parking space is typically legally restricted to being used for car parking. A typical planning permit condition is that 'parking areas and access lanes must be kept available for these purposes at all times' and that 'the car parking allocation as designated on the endorsed plan ... must be complied with at all times and to the satisfaction of the Responsible Authority' (*Green v Hobsons Bay CC* 2013, np). Car parking is thus made an integral part – physically and legally – of what constitutes an allowed dwelling.

As a result of such linkages, car parking factors into housing markets in both direct and indirect ways (Guo and Ren 2013; McDonnell *et al.* 2010; Manville 2013; Marsden 2006; Stubbs 2002). These include: occupying space; directly competing for surface-level space; adding to construction costs in basement or multi-level parking; reducing housing supply through trade-offs; and being priced into minimum housing space/footprint and, as a result, into minimum housing costs. Mitigating factors in the relationship between parking and housing markets include: local prices of housing and land; accessibility; residential self-selection; and minimum parking policies. The first three constitute variances in the underlying demand for housing and for parking, with both housing and parking demand influenced by spatial differences in the market for urban land.

Hedonic pricing studies indicate that off-street parking adds 10–12.5 per cent to dwelling costs (Litman 2006), although Stubbs (2002) emphasised the complexity of markets for residential parking. These include compromises forced by higher housing prices in high-demand inner areas. Where minimum parking policies apply, it is difficult to measure parking as a component of housing price. The clearest way in which parking policies are thought to distort housing cost is by increasing minimum housing costs in high-cost, high-demand, accessible areas where the underlying demand for parking space is elastic – where households would purchase more or less parking depending on their preferences and ability to pay (McDonnell *et al.* 2010; Manville 2013).

Major policy reforms in world cities – London, New York, Los Angeles – have sought to improve transport and affordability outcomes by removing minimum parking requirements (Guo and Ren 2013; Li and Guo 2014; McDonnell *et al.* 2010; Manville 2013; Manville *et al.* 2013; Wilson 2013).

Guo and Ren (2013) found that following the removal of minimum residential parking requirements across Greater London in 2004, the amount of residential parking provided reduced to 52 per cent of the prior regulatory requirement. Plans are in place to wind back parking requirements in New York City as part of zoning reforms for affordable housing (*New York Times*, 14 September 2014).

Although minimum ratios vary, few Australian local governments allow new housing without parking – the centres of Melbourne and Sydney are among the few exceptions. Traditional approaches to car parking persist in Australia partly because of vocal public opposition to changes to the status quo, with existing residents fearing loss of already competed-for on-street parking space and expressing doubt over transport alternatives (Cook *et al.* 2013; Taylor 2014). In Melbourne, the perceived loss of parking space associated with housing development prompted campaigning to 'SAVE this car park', and 'fight the towers! Or kiss your car park goodbye' (Taylor 2014). The prospect of allowing apartments near train stations to be built without parking was reported by Sydney tabloids as signalling a 'welcome to the third world' and 'car killing crusade' (*Daily Telegraph*, 25 September 2014), with the politicians involved depicted as 50-foot high monstrosities devouring cars. This climate of vigorous contestation creates a complex problem for local planning decisions, and moves to reduce statutory parking requirements have been tentative and contentious – as the case studies demonstrate.

Can you buy a house without car parking? Melbourne housing sales data

Car parking is therefore produced as a basic part not only of housing, but of housing markets. Except for housing built well before the advent of cars, it is comparatively difficult to purchase Australian housing without parking. Parking and housing are thus bound together in ways that, this chapter argues, become more apparent when attempting to build or to buy housing without or with less parking. This section asks: to what extent is it possible to purchase housing without car parking? It explores this question using data on housing sales for the Melbourne region.

The main data source is the Australian Property Monitors (APM) data set of property-level sold properties for Victoria (http://data.aurin.org.au/dataset/apm-generalsold-vicnoadrs-na), available through the Australian Urban Research Infrastructure Network (AURIN) portal. This data set contains information on the price and attributes of residential properties sold in the state. However, a rich, spatially disaggregated data set it is not necessarily representative of all property sales. APM data are compiled from 'real estate advertising, auctions, government and semi-government agencies, real estate agents and APM's own researchers' (AURIN 2005). The APM data set contains information on property characteristics, including the listed number of parking spaces.

For the analysis, one quarter – January to March 2015 – of property sales for the Melbourne statistical region have been extracted. Two sets of data are used – houses (detached houses) and units/apartments (other). These classifications are as applied by APM. There were 18,591 Melbourne house sales recorded in the APM data for quarter 1 of 2015; and 4,576 unit/apartment sales.

Figure 5.1 shows the share of detached house sales (of 18,591 records) for quarter 1, 2015 in Melbourne by the number of car parking spaces listed in the sale records. Nearly half (8,748 or 47 per cent) of houses sold across the city included two parking spaces. The next largest grouping (3,726 sales or 20 per cent of all houses sold) had one parking space. Significant portions of house sales also had three (6 per cent) or four or more (7 per cent) parking spaces listed. Thus the remaining 20 per cent (3,678 sales) of houses were listed as sold without parking.

Considering sales of houses without parking as a proportion of houses sold by suburb, the predominant suburbs include mostly older inner city suburbs: Abbotsford (17 of 21 sales); Princess Hill (4 of 6 sales); Albert Park (17 of 27 sales); Carlton North (11 of 18 sales); Fitzroy (11 of 19 sales); and Carlton (13 of 23 sales). To provide some validation of the APM parking

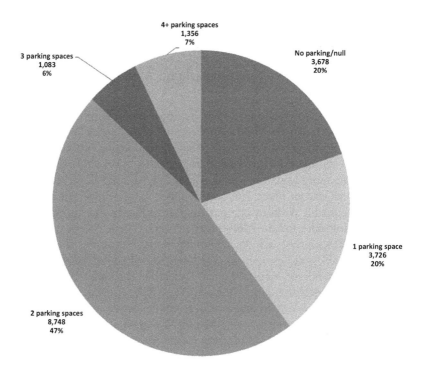

Figure 5.1 Detached house sales, Melbourne, quarter 1, 2015: by number of car parking spaces.

data for house sales, the first 100 houses listed for sale in the Melbourne region as at July 2015 were reviewed on property sales website Domain. com.au. Of these 100 sales, 35 per cent were essentially (for the purposes of this chapter) miscoded: typically in outer areas, where the site had open-air parking space but not an undercover garage, these were listed as not having parking. A further 49 per cent were period-era houses (mostly listed as 'Victorian terrace') predating private motor vehicles and therefore predating off-street parking requirements.

Sales for semi-detached and apartment housing tell a somewhat different story. The APM records for the 4,576 unit and apartment sales in Melbourne in quarter 1 of 2015 are shown at Figure 5.2 by the number of listed parking spaces. Of these, over half (2,499 or 55 per cent) had one parking space; and 851 (19 per cent) had two parking spaces. Only small numbers had larger numbers of parking spaces, leaving the remainder, 1,179 or 26 per cent of unit/apartment sales, with no car parking listed. As indicated in Figure 5.3, by far the largest share of apartments sold without parking were in the suburb of Melbourne – meaning the Melbourne Central Business District (CBD) (201

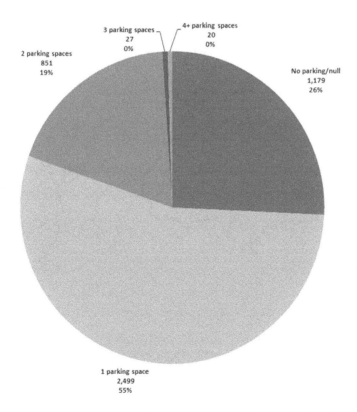

Figure 5.2 Unit/apartment sales, Melbourne, quarter 1, 2015: by number of car parking spaces.

Figure 5.3 Unit/apartment sales, Melbourne, quarter 1, 2015: number without car parking by suburb.

sales or 17 per cent of the total apartments sold without parking listed). Over half (53 per cent) of units/apartments sold in the CBD did not have parking. Other significant numbers of apartments without parking were in Carlton (92 sales); South Yarra (83); St Kilda (57); Southbank (44); and Brunswick (27).

Melbourne City Council (MCC) – the municipal authority for the Melbourne CBD, Carlton, Southbank and Docklands – does not apply minimum parking ratios for development including housing. Instead, MCC applies a 'parking overlay' to remove minimum rates and, in the central areas, caps maximum rates of parking (generally to two spaces per dwelling). At least 30 per cent of units and apartments sold across Melbourne without car parking were sold in this central core of the city, without minimum parking policies. This likely reflects a combination of high demand for housing in these inner-city locations, very high relative public transport accessibility and pressure for space in suburbs where a parking space (around 20 square metres) is significant relative to apartment sizes (40–50 square metres).

Further insight is given by considering the first 100 apartments listed for sale in the Melbourne region as at July 2015 on property sales website domain.com.au. Of these, 20 per cent were limited occupancy apartments: student-only, retirement-only or serviced apartments for investment only. These have limited mortgage finance options and are not generally available to owner-occupiers. Based on the sampled internet listings, a further 20 per cent of units/apartments sold without parking were older, pre- or inter-war 'art deco' flats. Lewis (1999) details how early higher-density dwellings in Melbourne were built primarily in the inner east.

Overall, over half (52 per cent) of all 23,167 dwellings (low-, medium- and high-density) sold in Melbourne in the sample period (quarter 1, 2015) had two or more parking spaces sold with them. A further 27 per cent had one parking space. This left 4,857 (21 per cent) of dwellings sold without parking. A sample validation of this latter data suggested that most dwellings listed as having been sold without car parking were houses in outer areas without undercover garages; older inner-city housing pre-dating cars; and apartments sold in the immediate CBD and surrounds where minimum parking requirements are not applied, with the latter including dedicated student or investment housing.

The data also suggest new housing is increasingly sold without parking in inner- and middle-ring suburbs. This accords with Taylor's (2014) review of planning appeals for Melbourne, which found development pressure for such housing – without or with less than statutory levels of parking – exists, often after applying for a planning 'waiver'. Such applications very frequently come into conflict within the development approval process. One such location is Brunswick, in the inner north of Melbourne in the LGA of Moreland. Brunswick is a location for considerable conflict over urban change and consolidation (Taylor 2013; Woodcock *et al.* 2011), and the location of the two case study developments discussed next.

Can you build housing without parking? 'The Commons' and 'Nightingale'

Although development models and statutory requirements tend to discourage it, some Australian higher-density housing developments have been allowed to be built with no or close to no on-site parking provision. Primarily these are in the CBD, or in dedicated student housing not available for owner occupation. Fewer are infill developments in inner suburbs.

'The Commons' is a development of 24 apartments, two 'artist spaces', one café and one retail space. It is situated in Brunswick, in the inner northern suburbs of Melbourne, within the LGA of Moreland. Critically to the project's approval, the site is adjacent to a train station (Anstey) as well as to a major bicycle route, and to other public transport (trams) and services. The project was approved by the Council in September 2011 and was completed in late 2013. 'The Commons' had no on-site car parking spaces.

Promotional material from developer Small Giants – a grouping of companies seeking 'to create, support, nurture and empower businesses that are shifting us to a more socially equitable and environmentally sustainable world' (Small Giants 2015) – sees 'The Commons' described as a 'vertical eco village' (Small Giants 2013). A range of sustainability features are listed, contributing to its eight-star energy rating. In promotional material the developer describes the absence of car parking as follows:

> The proximity of The Commons to public transport is critical to the overall sustainability of the development. Uniquely, there are no car spaces allocated to the residences of The Commons – residents share the 65 bike spaces located in a secure garage and share the GoGet car parked directly out the front of the building in Florence Street. Being only 6km from the CBD and a short walk to the Sydney Road shops, the luxury of a personal car space is not critical to enjoying the lifestyle that The Commons presents.
>
> (Small Giants 2013)

'The Commons' was directly cited, in 2014, in a subsequent application for a similar project on the same street, by the same developers and architects. The 2014 application was for a five-storey development of 20 dwellings and two retail spaces, branded as 'The Nightingale'. The proposal was approved by council in March 2015 (Moreland City Council 2015).

Victoria has relatively broad mechanisms available for third party objection and appeal rights (TPOAR) (Clinch 2006; Cook *et al.* 2012; Willey 2006). Infill housing developments are a focus for third-party objection. In the context of TPOAR, fear of competition for parking space has been shown to be a major factor in planning conflict over intensifying cities (Taylor 2014). It was therefore a showcase statistic for the Nightingale application that the proposal received few (three) objections and 177 letters of support. Letters of support are unaccounted for in Victorian statutory approval systems, hence a local councillor quoted as saying: 'We didn't have a column [in our internal filing system] for tallying the number of supporters' (*Architecture and Design* 2015).

The council report (Moreland City Council 2015) notes that, had statutory parking requirements been applied to The Nightingale application, 29 parking spaces (24 for dwellings and five for retail space) would have been required. What is not explicitly stated is that meeting the statutory parking requirement would have required an additional approximately 580 square metres of floor space. The calculation is misleading, however, in that omitting car parking was fundamental to the way both of the developments were conceived. The developments were marketed as new investments in non-car-based transport and as new models of housing: a critical factor in the reaction of the local authority.

The burdens of proof: reconceiving housing without parking

Both 'The Commons' and 'Nightingale' developments needed to demonstrate various characteristics in order to attain development approval without on-site car parking. Both were dependent on a council 'waiver' of standard parking requirements (although had this not been obtained, attaining the same waiver from the planning appeal authority, VCAT, would have been the likely if not guaranteed recourse). Local council support was influenced by the attitudes of potential (aspiring) residents in the buildings. In effect, the burden of proof was to show the sites were extraordinary.

Council support

Planning regulations are deeply involved in the supply of car parking: through 'predict and provide' minimum parking policies, legacies of twentieth-century zoning approaches emphasising separated land uses connected by private car. The two projects necessitated a waiver of standard Victorian car parking requirements. Considerations in assessing waivers include whether a market exists for lower car dependence in the location; the availability of other transport options; local traffic management; and characteristics of likely residents. These reflect similar factors in the literature on complex trade-offs between location, housing costs and car ownership (Stubbs 2002). Council support was critical to the approval of both developments. The case studies asserted, to vetting authorities, the validity of housing models that re-linked housing to alternative transport systems.

Housing developers

The developers and architects of the two sites publicly espoused differing finance models and the social ethics underlying the design and principles of the developments. While this appears to contribute to the popularity of the projects, it is notable that the motives of development proponents are not normally a valid consideration in planning approval. Implicitly, requirements for parking may function as a punitive tax on less altruistically motivated housing development proponents.

Consumer demand

Both developments carried the burden of proof to show that demand for housing without parking existed. This extended to showing how new residents would be required to pay for, and provide proof of, alternative transport habits: further illustrating the integration of housing with transport systems. Building on 'The Commons', 'The Nightingale' was able to specifically cite support from prospective owners – a fact itself indicative of the 'waiting list' and audition-type process required to access the buildings: 'Some of the letters of support are from prospective owners who state that they will not require a car if living in this building' (Moreland City Council 2015).

   The burden of proof appears to fall on specific developments to prove that transport shifts are possible. In the case study sites, relatively detailed proof of the personal preferences of residents was given in order to justify a shift away from default parking provision. As well as leveraging strong underlying demand for the developments, both developments were required to submit a 'Green Travel Plan' as part of formal planning approval. This detailed bicycle spaces, car-share arrangements and a 'sustainable transport levy' allocating contributions per bedroom to public transport cards, car-share membership and bicycle servicing. Again it is notable that the personal habits of residents are not, notionally, valid planning considerations. Housing production systems are positioned as a veto of socially plausible ways of occupying housing and cities.

Location

The two developments were justified by being essentially on top of a railway station and bicycle path, and very close to other high-standard transport infrastructure and services. The Australian housing market reflects high demand for areas with higher public transport accessibility. Data from the 2011 Census also indicate rates of car use are much lower, and rates of car ownership somewhat lower, in inner urban areas well served by public transport. Taking Victoria as an example, in inner-city areas car ownership is still high (over 80 per cent), but car usage is lower. In the CBD – where parking restrictions do not exist – car ownership is lower (ABS 2011). In outer areas multiple car ownership is common and essentially all journeys are made by car. Perkins *et al.* (2009) also found that car use was considerably higher in the outer suburbs. Ultimately, without being placed in the midst of the high-quality transport infrastructure of past decades – making life without car use broadly possible – the development sites could not have justified a waiver of parking requirements. These kinds of decisions both reflect and reinforce inequity in the spatial distribution of infrastructure in Australian cities.

Design and sustainability

To attain council and consumer support, the two developments were widely felt to have exceeded the quality and sustainability standards of other housing developments. Hence: 'The main aspects raised by supporters of the proposal are high quality architecture, sustainability and affordability' (Moreland City Council 2015). Some of the evidence used in support of the two sites was from potential homebuyers aspiring to better-quality housing that, normally bundled with parking, would otherwise be unaffordable to them. Hence a prospective resident submitted that: 'I couldn't afford anything else of that quality as a single person' (*Architecture and Design* 2015).

What is broadly supported by the development approval is a market choice: allowing for housing consumers to substitute parking for other housing qualities. This is first carefully screened as an alternative configuration of home and transport. Such choices are only allowed on a case-by-case basis: only where the design and sustainability of a development is held to be extraordinarily high is residential car parking allowed to be left out. While meeting planning objectives, this sequence reinforces the prevalence of parking ratios in comparison to other comparatively weak policy levers for housing quality and sustainability in Australia. Parking requirements function perhaps as a last bastion of prescriptive regulation.

Other people's parking

As an inner-city suburb, neighbours in areas surrounding the two sites have existing use rights to park on the street. By contrast, residents of new higher-density developments – as in most developments across Melbourne – are specifically precluded from on-street residential parking rights. Nonetheless, and as in other similar developments, the prospect of apartment residents competing for on-street parking was a point of contention. Fears about inadequate parking are a significant issue in over half of Victoria's thousands of annual planning appeals, and over 80% of appeals that involve resident third-party objectors (Taylor 2014).

In the case study sites a key concern was that new residents would, despite the principles of the developments, nonetheless own cars and park them on the street. The three objections to the Nightingale concerned inadequate parking. Interestingly, the suspect transport habits of Commons residents were raised in objections to the new development:

> Two objections have been received on grounds that the proposed reduction of car parking will impact the availability of car parking in

the street. They have raised concerns that occupiers of the development at 7 – 9 Florence Street (which also does not provide car parking on site) own cars and park them in the street and surrounding area.

(Moreland City Council 2015)

In response, the council emphasised on-street parking regulations:

Whether some occupiers of 7 – 9 Florence Street have cars should not preclude the proposed development from being permitted to have a parking reduction. This type of development actively discourages car ownership, but it cannot prohibit it. Nevertheless, occupiers will only be able to park in the street in accordance with parking regulations. Owners and/or occupiers of the premises will not be eligible for any Council parking permits to allow for on street parking. This is noted in the recommendation.

(Moreland City Council 2015)

A broadly similar but lower-density site – the WestWyck 'eco village' in West Brunswick – was earlier the subject of two VCAT appeals to attain development approval. The main issue in that case was car parking, with concerns raised by neighbours to the site (*Brelis & Ors v Moreland CC [2013]*; *Brellis & Ors v Moreland [2011]*). To defend against 'inadequate residential car parking provision' (in this case reduced, not removed) the WestWyck developers were likewise required to emphasise the 'eco village' nature of the development, its location and the expected practices of residents.

These conflicts over on-street parking emphasise two issues. First, that there is very little, if any, evidence of the actual transport practices of residents of low- or no-parking developments. Second, they highlight the importance of on-street parking rights and their allocation and enforcement. The availability, price and enforcement of on-street parking rights are critical factors in the parking literature (Guo 2013; Shoup 2005). Newer parking management approaches tend to advocate open pricing for on-street and off-street parking (Litman 2006). As it stands, the burden of proof is on the housing production system to predict the future car ownership of residents. This does not allow for flexibility in car ownership, and also reinforces the asymmetry of access rights to public parking. Introducing any pricing to parking is, however, politically extremely difficult: the existing metering of on-street parking in the CBD (itself the outcome of decades of earlier conflict – see Davison and Yelland 2004) is perhaps one reason why reduced off-street parking requirements exist there.

Implications

A growing international literature suggests that car parking policy plays an important role in balancing housing and other urban goals. At present there is limited evidence concerning the costs and benefits of residential parking policy in Australian cities. Little is known about the actual transport practices of people living in developments without or with reduced or unbundled parking; about the extent to which Australian residents trade-off parking for other qualities; and about how different developers respond in practice. Instead, housing provision relies on a normative conception of transport and housing practices, based on embedded twentieth-century practices.

Minimum parking policies are widespread, despite evidence of the complexity of underlying demand for links between transport and housing. This chapter demonstrated the comparative rarity of housing without parking in Melbourne, and its spatial concentration. A case study of two adjoining Melbourne housing developments built without on-site car parking was then given. In order to allow the supply of housing without car parking on the sites, a series of overlapping linkages needed to be purposefully untangled, and a considerable array of social, economic and cultural capital applied. A remarketing of housing had to be presented and vetted: without this and its requisite cultural capital, the default is to include car parking in new housing.

High, and rising, demand for housing and land in the inner north of Melbourne underscored the effort put into the two case study sites as well as the demand for apartments in them. Ultimately the two developments are priced similar to or above other apartments in the area. This is in part through the complexity of demand for housing and parking. Stubbs (2002), in a UK study, found that even households without cars typically preferred to purchase more parking – because of its resale value to others. To overcome concern around the loss of value often associated with removing parking, the two sites were not branded simply as cheaper, but as offering other niche values not otherwise available. The absence of car parking was specifically highlighted as a design feature of 'The Commons', and as a means of reducing apartment costs: 'Through removing the typical ground level car park the project both promotes more environmentally friendly transport options and significantly reduces the cost per apartment' (*Architecture and Design* 2015).

Although reducing costs, this process of justifying the sites' characteristics for planning approval fed into an exclusivity of the two case study sites. To access the apartments, waiting lists and an (unknown) selection process are involved. Many of the 177 submissions to support The Nightingale were from 'prospective' residents, far exceeding the 20 dwellings to be built. Perhaps for this reason, 'equitable development opportunities' were raised in objections (Moreland City Council 2015).

The effort applied in the case study sites and the exclusivity of them is indicative of the extent to which the default of housing in Australian cities is to be bound with car parking. Implicitly, lower-quality developments must also have car parking. This suggests two basic choices: low-quality, poorly designed, high-cost housing with car parking. Or the same housing, only what embedded practices imply must be even worse – that is, without parking.

However, some signs point to pressure on parking space and on these embedded twentieth-century linkages between housing, parking and cars. The legal tie of designated space to parking was tested recently in Northcote, Melbourne, when residents protested against a council enforcement order stopping them using parking space at an apartment building as a garden (*Herald Sun*, 25 March 2015). A growing grey literature reconsiders the physical space for parking as a resource with which to reimagine high-value urban space. Possibilities for change are suggested by initiatives like Parklet – temporarily replacing parking with green public space; by the potential of longer-term improvements to design quality of parking space (Ben-Joseph 2012; Wilson 2013); and by design innovation – as in 'Dormant Car Park Converted to Luxury Apartments' (*The Age*, 16 February 2015). 'Peak car' describes a growing (though still small) proportion of younger people driving less or starting driving later (Delbosc and Currie 2013). Car sharing, ride sharing and other technologies promise further uncertainty over future car ownership and parking (Metz 2013).

Most of these shifts are around the edges of dominant car-based transport patterns in Australia, and vary greatly by location. Location-specific conflicts and shifts like those described, combined with nascent scholarly and policy criticisms of mandatory parking requirements, may yet see a future market reconfiguration away from established packages of housing, parking and private car ownership. It may yet be possible for the car, and its space, to move out of home.

Postscript

In late October 2015 the Victorian Civil and Administrative Tribunal (VCAT) overturned Moreland City Council's decision to allow no parking provision at the Nightingale site. In a ruling ordering that the site must provide car parking, the VCAT member stated that although the site offered multiple transport options, none were 'as convenient as private car ownership' (*VCAT 2015 Chaucer Enterprises Pty Ltd v Moreland CC*). The VCAT case had been instigated by a neighbouring developer who had applied for reduced car parking but been denied. The decision has attracted a high level of media and commentator debate, with scrutiny both of parking regulations, case-by-case waivers and of reduced-parking developments.

References

Australian Bureau of Statistics. 2011. *Census of Population and Housing 2011: Custom Table – Number of Motor Vehicles and Dwelling Structure*. ABS Table Builder. Retrieved July 2012 from www.abs.gov.au/websitedbs/censushome.nsf/home/tablebuilder

Australian Bureau of Statistics 2013. *Motor Vehicle Census. (Cat. 9309.0).* Retrieved July 2012 from www.abs.gov.au

Architecture and Design 2015. Nightingale takes off: Melbourne architects' development approved. Retrieved 10 July 2015 from www.architectureanddesign.com.au/features/features-articles/nightingale-takes-off-melbourne-architects-develop.

Australian Bureau of Statistics (ABS). 2012. *Environmental Issues: Waste Management, Transport and Motor Vehicle Usage*, March, Cat. 4602.0.55.002.

Australian Urban Research Infrastructure Network (AURIN). 2015. Australian property monitors data. Retrived 15 September 2015 from http://data.aurin.org.au/organization/about/apm

Bagley, M.N. and Mokhtarian, P.L. (2002). The impact of residential neighborhood type on travel behavior: a structural equations modeling approach. *Annals of Regional Science*, 36(2): 279–297.

Ball, M. 2003. Markets and the structure of the housebuilding industry: an international perspective. *Urban Studies* 40(5–6): 897–916.

Barter, P. 2011. Parking requirements in some major Asian cities. *Transportation Research Record* 2245(2011): 79–86.

Ben-Joseph, E. 2012. *ReThinking a Lot: The Design and Culture of Parking*. MIT Press, Cambridge, MA.

Brelis & Ors v Moreland CC. 2011. AUSTLII database of VCAT decisions. Retrieved July 2012 from www.austlii.edu.au/cgi-bin/sinodisp/au/cases/vic/VCAT/2011/769

Brelis & Ors v Moreland CC. 2013. AUSTLII database of VCAT decisions. Retrieved July 2012 from www.austlii.edu.au/cgi-bin/sinodisp/au/cases/vic/VCAT/2013/33

Bunker, R., Gleeson, B., Holloway, D. and Randolph, B. 2002. The local impacts of urban consolidation in Sydney. *Urban Policy and Research* 20(2): 143–168.

Cao, J. and Cao, X. 2014. The impacts of LRT, neighbourhood characteristics, and self-selection on auto ownership: evidence from Minneapolis-St. Paul. *Urban Studies* 51(10): 2068–2087.

Clinch, J.P. (2006). Third party rights of appeal: enhancing democracy or hindering progress? *Planning Theory & Practice*, 7(3): 327–350.

Cook, N., Taylor, E. and Hurley, J. 2013. At home with strategic planning: reconciling resident attachments to home with policies of residential densification. *Australian Planner* 50(2): 130–137.

Daily Telegraph. 2014. 'Liberal MPs smash Pru Goward for joining Clover Moore's car-killing crusade', 25 September.

Dalton, T., Hurley, J., Gharaie, E., Wakefield, R. and Horne, R. 2013. *Australian Suburban House Building: Industry Organisation, Practices and Constraints*, AHURI Final Report No. 213. AHURI, Melbourne.

Davison, G. and Yelland, S. 2004. *Car Wars: How the Car Won Our Hearts and Conquered Our Cities*. Allen & Unwin, Crows Nest.

Delbosc, A. and Currie, G. 2013. Causes of Youth Licensing Decline. *Transport Reviews* 33 (3): 271–290.

Dodson, J. and Sipe, N. 2007. Oil vulnerability in the Australian city: assessing socioeconomic risks from higher urban fuel prices. *Urban Studies* 44(1): 37–62.

Easthope, H. and Randolph, B. 2009. Governing the compact city: the challenges of apartment living in Sydney, Australia. *Housing Studies* 24(2): 243–259.

Featherstone, M. 2004. Automobilities: an introduction. *Theory, Culture & Society* 21(4–5): 1–24.

Fuller, G. 2015. In the garage. *Angelaki* 20(1): 125–136.

Gardner, B. and Abraham, C. 2008. Psychological correlates of car use: a meta-analysis. *Transportation Research Part F: Traffic Psychology and Behaviour* 11(4): 300–311.

Godelier, E. 2007. 'Do you have a garage?' Discussion of some myths about entrepreneurship. *Business and Economic History Online* 5 (2007): 1–20.

Goodman, R., Buxton, M., Chetri, P., Taylor, E. and Wood, G. 2010. *Planning and the Characteristics of Housing Supply in Melbourne*, AHURI Final Report No. 157. AHURI, Melbourne.

Green v Hobsons Bay CC. 2013. AUSTLII database of VCAT decisions. Retrieved July 2012 from www.austlii.edu.au/cgi-bin/sinodisp/au/cases/vic/VCAT/2013/2091

Guo, Z. 2013. Residential street parking and car ownership. *Journal of the American Planning Association* 79(1): 32–48.

Guo, Z. and Ren, S. 2013. From minimum to maximum: impact of the London parking reform on residential parking supply from 2004 to 2010. *Urban Studies* 50(6): 1183–1200.

Gurran, N. 2008. Affordable housing: a dilemma for metropolitan planning? *Urban Policy and Research* 26(1): 101–110.

Hagman, O. 2006. Morning queues and parking problems. *Mobilities* 1(1). 63–74.

Hall, T. 2008. Where have all the gardens gone? *Australian Planner* 45(1): 30–37.

Herald Sun. 2015. Group's bid to keep community garden in Northcote car park, 25 March. Retrieved July 2012 from www.heraldsun.com.au/leader/north/groups-bid-to-keep-community-garden-in-northcote-car-park/story-fnglenug-1227277778200

Hess, P.M. 2008. Fronts and backs: the use of streets, yards, and alleys in Toronto-area new urbanist neighborhoods. *Journal of Planning Education and Research* 28(2): 196–212.

Jakle, J. and Sculle, K. 2004. *Lots of Parking: Land Use in Car Culture*. University of Virginia Press, Charlottesville.

Kauppila, P. 2005. The sound of the suburbs: a case study of three garage bands in San Jose, California during the 1960s. *Popular Music and Society* 28(3): 391–405.

Kent, J.L. 2014. Still feeling the car: the role of comfort. *Mobilities* 10(5): 726–747.

Lewis, M. 1999. *Suburban Backlash: The Battle for the World's Most Liveable City*. Bloomings Books, Hawthorn.

Li, F. and Guo, Z. 2014. Do parking standards matter? Evaluating the London parking reform with a matched-pair approach. *Transportation Research Part A: Policy and Practice* 67: 352–365.

Litman, T. 2006. *Parking Management Best Practices*. American Planning Association, Chicago.

McDonnell, S., Madar, J. and Been, V. 2010. Minimum parking requirements and housing affordability in New York City. *Housing Policy Debate* 21(1): 45–68.

Manor Lakes. 2015. *Manor Lakes Urban Design Guide*. Retrieved 10 July 2015 from www.manorlakes.com.au/sites/default/files/Manor-Urban-Design-Guides-Stage-118A.pdf

Manville, M. 2013. Parking requirements and housing development. *Journal of the American Planning Association* 79(1): 49–66.

Manville, M., Beata, A. and Shoup, D. 2013. Turning housing into driving: parking requirements and density in Los Angeles and New York. *Housing Policy Debate* 23(2): 350–375.

Marsden, G. 2006. The evidence base for parking policies: a review. *Transport Policy* 13(6): 447–457.

Marusek, S. 2011. *Politics of Parking: Rights, Identity, and Property*. Ashgate, Burlington, VT.

Mees, P. and Groenhart, L. 2012. *Transport Policy at the Crossroads: Travel to Work in Australian Capital Cities 1976–2011*. RMIT, Melbourne. Retrieved July 2012 from http://mams.rmit.edu.au/ov14prh13lps1.pdf

Metz, D. 2013. Peak car and beyond: the fourth era of travel. *Transport Reviews* 33(3): 255–270.

Mingardo, G., van Wee, B. and Rye, T. 2015. Urban parking policy in Europe: a conceptualization of past and possible future trends. *Transportation Research Part A: Policy and Practice* 74(2015): 268–281.

Moreland City Council. 2015. *Urban Planning Committee Agenda*, 25 February 2015.

Motte-Baumvol, B., Massot, M.-H. and Byrd, A.M. (2010). Escaping car dependence in the outer suburbs of Paris. *Urban Studies* 47(3): 604–619.

Moudon, A. 1992. The evolution of twentieth-century residential forms: an American case study. In J.W.R. Whitehead and P.J. Larkham (eds) *Urban Landscapes: An International Perspective*. Routledge, London.

Mukhija, V. and Shoup, D. 2006. Quantity versus quality in off-street parking requirements. *Journal of the American Planning Association* 72(3): 296–308.

Nasar, J., Evans-Cowley, J. and Mantero, V. 2007. McMansions: the extent and regulation of super-sized houses. *Journal of Urban Design* 12(3), 339–358.

New York Times. 2014. Trading parking lots for affordable housing, 14 September.

Pandhe, A. and March, A. 2012. Parking availability influences on travel mode: Melbourne CBD. *Australian Planner* 49(2): 161–171.

Perkins, A., Hamnett, S., Pullen, S., Zito, R. and Trebilcock, D. 2009. Transport, housing and urban form: the life cycle energy consumption and emissions of city centre apartments compared with suburban dwellings. *Urban Policy and Research* 27(4): 377–396.

Pierce, G. and Shoup, D. 2013. Getting the prices right. *Journal of the American Planning Association* 79(1): 67–81.

Randolph, B. (2006). Delivering the compact city in Australia: current trends and future implications. *Urban Policy and Research*, 24(4): 473–490.

Ruming, K. and Houston, D. 2013. Enacting planning borders: consolidation and resistance in Ku-ring-gai, Sydney. *Australian Planner* 50(2): 123–129.

Searle, G. and Filion, P. 2011. Planning context and urban intensification outcomes. *Urban Studies* 48(7): 1419–1438.

Shoup, D. 2005. *The High Cost of Free Parking*. Planners Press APA, Chicago.

Small Giants. 2013. A new eco-village is born – the Commons is complete. Retrieved 10 July 2015 from www.smallgiants.com.au/a-new-eco-village-is-born-the-commons-is-complete.

Small Giants. 2015. Our purpose. Retrieved 10 July 2015 from www.smallgiants.com.au/our-purpose.

Stubbs, M. 2002. Car parking and residential development: sustainability, design and planning policy, and public perceptions of parking provision. *Journal of Urban Design* 7(2): 213–237.

Taylor, E. 2013. Do house values influence resistance to development? A spatial analysis of planning objection and appeals in Melbourne. *Urban Policy and Research* 31(1): 5–26.

Taylor, E. 2014. 'Fight the towers! Or kiss your car park goodbye': how often do residents assert car parking rights in Melbourne planning appeals? *Planning Theory and Practice* 15(3): 328–348.

The Age. 2015a. Dormant car park converted to luxury apartments. 16 February.

The Age. 2015b. South Yarra residential parking permits under threat, 2 September.
Vanderbilt, T. 2008. *Traffic: Why We Drive the Way We Do*. Allen Lane, London.
Willey, S. (2006). Planning appeals: are third party rights legitimate? The case study of Victoria, Australia. *Urban Policy and Research*, 24(3): 369–389.
Wilson, R.W. 2013. *Parking Reform Made Easy*. Island Press, Washington, DC.
Woodcock, I., Dovey, K., Wollan, S. and Robertson, I. 2011. Speculation and resistance: constraints on compact city policy implementation in Melbourne. *Urban Policy and Research* 29(4): 343–362.
Zhao, J., Peters, A. and Rickwood, P. (2013). Effect of raising fuel price on reduction in household transport greenhouse gas emissions: a sydney case study. *2013 State of Australian Cities (SOAC) Conference*, Sydney.

Part II

Housing/home and worlds of nature

Andrew Gorman-Murray, *The External 2*, 2015.

Andrew Gorman-Murray, *The Internal* 5, 2015.

Andrew Gorman-Murray, *The Internal 8*, 2015.

Andrew Gorman-Murray, *The Interstitial 3*, 2015.

6 Secure in the privacy of your own nature

Political ontology, urban nature and home ownership in Australia

Aidan Davison

Modernist accounts of cities as autonomous, tightly bound spaces from which nature has been expelled have been longstanding and widespread (Williams 1973). This distinction belongs to a network of ontological dualisms that set colonial-capitalist projects of modern progress apart from the worlds they sought to master (Plumwood 1993). This production of cities as unnatural space is now under challenge as part of conceptual and material dynamics in late-modern societies that are unsettling divisions between nature and culture (Latour 2004).

Long avoided as an oxymoron, the figure of urban nature has enabled new forms of dialogue across the social and natural sciences over the last three decades (Braun 2005; Heynen 2014; Wolch 2007). In the process, cities are being reconfigured within regional and global geographies, calling into question the status of nonurban spaces and drawing attention to multiscalar webs of interaction and interdependence. This is to engage with urbanisation as a planetary phenomenon integral not just to capital accumulation, but to the future of life on earth. While now popular to conceive of as the Anthropocene, an epoch in which human action has planetary force, our time is more precisely understood as the Urbanocene, a prospect in which earthly possibilities are definitively urban (Gleeson 2015). And here the newly animated figure of urban nature takes on vital work, asserting that to declare our times urban is not thereby to declare them unnatural.

In this chapter I contribute to emerging research in more-than-human urbanism by offering a preliminary inquiry into the role of private home (and land) ownership in Australia in the shifting politics of urban nature.[1] Capitalist discourses and practices of home ownership have been centrally implicated in the political, cultural and material boundaries of home in Australia. I begin by critically engaging the literature of urban political ecology, identifying within it a need for a relational ontology of home ownership. I then sketch the contours of home ownership in the history of Australian modernity, exploring its significance in the ontological politics of urban nature. Of course, to suggest that home ownership is implicated in mutually constitutive relations between nature and society is to question received knowledge about many things, including cities. The modern dualism that underwrites discursive and material divisions between nature and city underwrites also the boundaries of the natural and the social sciences. This chapter thus contributes to the unbounding of research about housing and home by opening home ownership to inquiry about the politics that circulate in and through the production of urban natures.

Urban political ecology and home ownership

Urban political ecology is notable among scholarship of urban nature for trenchant inquiry into how and with what effect more-than-human realities participate in political economic systems. A 'first wave' of urban political ecology took shape in the early years of this millennium (Gandy 2012; Heynen 2014; Heynen *et al.* 2006; Keil 2003). Although eclectic in theory, method and intent, this endeavour shares the premise that cities are 'socio-ecological and political-economic processes' (Harvey 1996, 6) arising from the dialectical coproduction of nature and society. As a result, 'the material and symbolic, the natural and the cultural, the pristine and the urban are not dual and separate realities, but rather intertwined and inseparable aspects of the world we inhabit' (Keil 2003, 728). Grounded in neo-Marxian analysis, the first wave of this inquiry has explored how cities function within capitalist political economy to distribute money, resources, risks and power in systematically uneven ways.

Questions of land ownership have often been at the centre of political ecology studies in nonurban settings in the 'Global South' (McCarthy 2005). More widely, many have challenged the neoliberal assumption that private property ownership ensures environmental protection through enlightened self-interest (Merrill 2004; Meyer 2009). Yet analysis of property relations has been much less prominent in political ecology studies of modern societies, particularly in cities (McCarthy 2005).

One of relatively few urban political ecology studies to make home ownership explicit is Nik Heynen and Harold Perkins' (2005) investigation of urban tree distribution in Milwaukee. Conceiving urban environments as a consumption fund of fixed capital that encompasses 'urban ecological resources,' Heynen and Perkins (2005, 105–106) document a decline of public investment in Milwaukee's trees:

> Historically, many artifacts within the consumption fund, for instance parks and urban forests, have been underwritten and maintained publically. Current processes of neoliberalization, steeped in free-marketeering rhetoric, have increased the apparent legitimacy and inevitability of the right to own private property.

They argue that neoliberal capitalism is creating 'an increasingly uneven urban ecology in Milwaukee, relegating lower income groups to areas that lack trees' (Heynen and Perkins 2005, 111).

This argument reveals four limitations in the first wave of urban political ecology analysis. First, Heynen and Perkins (2005) uncritically accept the existence of a social consensus that trees are an urban ecological good, despite evidence that attitudes towards trees are subject to considerable cultural and individual variation (Kirkpatrick *et al.* 2013). This acceptance is based upon an unexamined scientific discourse – in this case about the intrinsic benefits of trees and evident in the claim that the loss of trees in Milwaukee 'contributes in some measure to the degradation of the global ecosphere' (Heynen and Perkins, 2005, 111) – that is treated sceptically in political ecology critique in nonmodern contexts. Second, these authors emulate while critiquing the reduction of trees to economic abstractions, subsuming tree agency within a reified agency of capital, with its 'habit of annihilating space with time' and 'never-ending quest for the expansion of value' (Heynen and Perkins 2005, 103). Third, they subsume urban residents within a homogenising and deterministic account of 'individual capitalists', a social identity with a prefigured 'reluctance to invest in infrastructures that renew the cycle of production and accumulation' (Heynen and Perkins 2005, 106). Fourth, having denatured trees as capital and objectified residents as capitalists, Heynen and Perkins offer no analysis of the lived relationships between trees and people in Milwaukee. Thus, despite noting that trees have 'contributed to the sense of place and standard of living' and that 'Milwaukeans have historically loved their trees; ... part of their shared identity' (Heynen and Perkins 2005, 108), this love is treated as undifferentiated background of little political relevance.

To the extent that it is acknowledged, private home ownership is presented in the first wave of urban political ecology as a political economic relationship that allocates unevenly the benefits of urban nature; and it is this. Yet home ownership is more than this, and more fundamental than this. Home ownership is bound up in the political activity of making and representing urban natures; in giving these realities diverse forms and meanings. As Nicholas Blomley (2015, 1) puts it, 'private property in land' is a 'relational phenomenon' that does nothing less than 'organise the world for us.' Home ownership does not merely express what it means to be at home in the world, it helps establish what it could mean to be at home in the world. Home ownership is part of what Kevin Grove (2009, 208) calls the '"cultural politics" of meaning in which the stakes are precisely what counts as "nature"'. We need, then, to investigate not just the distributional effects of any given urban nature, but to account for this given-ness itself. This is to pay attention to the way 'struggles over meanings and practices of nature and the city ... shape identities that make some forms of urban metabolisms possible while foreclosing others' (Grove 2009, 209).

Neglect of the role of home ownership in the making and the meaning of urban natures in the first wave of urban political ecology reflects the way in which its 'political economic roots ... delimit the possibilities for thinking through politics and gloss over the potentialities inherent in contested meanings of the non-human for radical politics' (Grove 2009, 208). Consider how, in welcoming the advent of urban political ecology, Neil Smith (2006, xiv) observes that urban nature

> seems far less visible, precisely because the arrangement of asphalt and concrete, water mains and garbage dumps, cars and subways seems so inimical to our intuitive sense of (external) nature. Whatever our analytical sophistication, the idea of nature as a contrivance still cuts deeply and sharply against our most engrained and peculiar prejudices.

Sharing this problematic appeal to 'the intuitive', many exponents of urban political ecology replace posters of pristine wilderness with those of a thoroughly technologised vista, celebrating artifice as humanity's natural habitat. In affirming technologised nature, however, this critique risks leaving the underlying structure of modern dualism intact, requiring that *either* nature *or* society be made subordinate to the other. It thereby narrows the filter through which nonhuman agencies are perceived and exaggerates the nonhuman relevance of urban boundaries. Although spatial scale is but one example of this narrowing, consider the microscopic city. This is a place in which each human gut, and many a carpet, rivals the Amazon for ecological complexity, and in which the ground beneath a skyscraper as much as beneath a park teems with nematodes, a form of life estimated to account for four in five animals on Earth. Or consider the macroscopic city, a place shot through with sublime interstellar forces – not least gravity and electromagnetic radiation – fundamental to our existence on a whirling planet whose surface is presently warming but whose molten depths continue, ever so slowly, to cool.

Smith's difficulty in articulating the presence of nonhuman things, beings and forces in the technological world of the city is widely shared and understandable. It is a difficulty sustained by the ongoing event of modern urbanisation. Masking of the more-than-human dimensions of modern cities has always been more than cognitive. Perception of cities as separate from nature is cause and effect in the material production of urban spaces in ways that exaggerate human agency, deny nonhuman agency and assert rigid boundaries between the two. The dialectic of instrumentalist mastery over nature and romanticist yearning for nature that Max Horkheimer and Theordore Adorno (1969) identified is materialised in the urbanisation process. Modern urban realities are the material performance of this dialectic, aligning with, and thereby validating, this quest. Rendered components of human designs, nonhuman constituents of modern cities are denatured, most often by being technologised as raw material or aestheticised as decoration.

Towards a political ontology of urban nature

Although it allows that the city rests upon a process of socio-natural metabolism, the first wave of urban political ecology has struggled to articulate nonhuman agencies in the making and the meaning of urban realities. Responding to this limitation, the last five years have seen the rise of a 'second wave' of this inquiry less given to determinism about historically unfolding structures, less restricted to city environments, more interested in more-than-human sources of human possibility, and more aware of the lived encounters that constitute urban natures (Angelo and Wachsmuth 2015; Heynen 2014). This version of urban political ecology is informed by relational ontologies indebted to poststructuralist critique, and associated with the posthumanism of Bruno Latour, Donna Haraway and others (Gabriel 2014; Gandy 2012). This is to work with an irreducible tension between materially embedded power relations and relationally emergent material realities; between hegemony and multiplicity, imperative and contingency. In resisting the temptation to resolve this tension through universal abstractions, this inquiry is sensitive to ontological difference, indeterminacy and plurality. This is not simply to acknowledge competing accounts of reality, but to be attuned to the ontological politics (Mol 1999) created by the enacting of different but coexisting realities. This is to allow that reality is comprehended only in acts of generative encounter; that ontology is first and foremost a practice of becoming (Davison 2001).

Emerging inquiry into the ontological politics of urban nature engages with urbanisation as a process of socio-natural becoming. This is an inherently open political process in which some human and natural possibilities are realised but not ordained, while others are suppressed but not foreclosed. The emergent and contingent character of this process arises as a function of its performative ground in more-than-human worlds of practice. This is not to deny the existence of enduring and relatively stable socio-natural regimes of power, such as capitalism. Because representations of reality are indivisible from encounters in more-than-human worlds of practice, powerful performative alignments are created between ways of making the world and ways of knowing the world. The alignment of urban realities and dualistic epistemologies described above is part of the relative stability of modern projects of technological progress. Yet this alignment is not historically destined. To endure, it must be constantly reproduced though the action of countless human and nonhuman agents. While these alignments constrain choice, they neither eliminate human and nonhuman autonomy, nor fully account for the unpredictability of complex and dynamic relations of socio-natural coproduction, nor transcend the particularities of geography and history.

It is within the second wave of urban political ecology that I situate my interest in private home ownership. Put ontologically, all modes of home occupation, including capitalist ownership, are part of an ongoing gathering together of a hospitable world within the vast, boundless field of existence that dwarfs all human comprehension and control. At once material and semiotic, rational and affective, this gathering together is the essence of the everyday practices of home ownership, or what Blomley (2013, 25) calls the 'reiterated ... performances of property', such as fence building, place-based identities, mortgage payments, gardening, narratives of privacy, house decoration, reticulated sewerage disposal and the rest. In conventional political terms, these performances demarcate public and private realms, allocate civic responsibilities and rights, establish social conditions of belonging and exclusion, confer economic advantages and disadvantages and distribute environmental goods and bads. Viewed from beyond humanist boundaries, these practices of becoming also influence the expression of human and nonhuman possibilities that, in turn, shape these worlds. This is to engage a posthumanist politics in which practices such as home ownership are recognised to involve choices about the making and the meaning of human and nonhuman natures.

In Australia, private home ownership is a hegemonic form of everyday life that does much to constitute the material arrangements, cultural meanings and political functions of nature, including urban natures. Yet this is also a constellation of practices whose function has long been to bound the private home in ways that mask its relationship to collective, social and natural, realities. The hegemony of this practice is not predestined nor uncontested. Indeed, this practice has been far from stable or unequivocal in modern Australia. In what follows, I investigate the historical role of home ownership in the dialectical production of nature and society, before considering the shifting nature of this role in the current quest for urban sustainability in Australia.

City, nature and modernity in Australia

As noted at the outset, a distinction between built environments and natural environments has underpinned the production of modern cities as spaces in which technology supersedes nature. In the nineteenth century, Australian urbanisation held an unfamiliar nature at bay, enabling fledgling colonies to graft European dreams onto a new world. Today, the resulting cities hold most of the population while their geographies of provision and waste ranged over the continent (and around the planet), with an Indigenous order perceived to be fading and hanging on in remote outposts. Indeed, 69 per cent of the population live in and around just six coastal cities, with these capital cities' populations increasing by 17 per cent in the decade to 2011, while the rest of the population grew by only 11 per cent (ABS 2012a).

Yet in the making of Australian modernity, experience of built environments cultivated yearning for natural environments. Faced with a stark choice between technological progress and the strangeness of the Australian continent, many made their home in ambivalent, hybrid landscapes in which they sought to claim the best and to avoid the worst of the city and of nature. Not for nothing was Australia the world's first thoroughly suburban nation (Davison 1995). These cities were suburban before they were urban, displaying a novel settlement structure explained poorly by categories inherited from European experience. From the first suburban boom of the 1880s, Australian desire for nature has been directed towards spaces of authenticity, innocence and grace understood to lie beyond the taint of the city. This was a desire for roots, a need for belonging, fed by the anxieties of inhabiting an alien land in the aftermath of invasion.

Wedded to an urban economy, but with memories of Europe's early industrial urban chaos fresh, an egalitarian cross-section of the population purchased land on the edge of cities, facing out from behind the safety of suburban hedges towards a geographical 'great beyond'. When first documented in the late 1880s, around half of all housing in the Australian colonies was occupied by owners (Butlin 1964, 259). The 1911 Census similarly recorded that 45 per cent of homes were owned outright, with a further 4 per cent being purchased (Williams 1984, 171). Relatively high disposable incomes combined with a relative lack of public revenue, infrastructure and services in the first century of Australian cities to ensure that private homes and gardens were vital sources of self-provision. By the early decades of the twentieth century, Australia boasted a large suburban peasantry engaged in food cultivation and processing, resource collection and waste disposal and building, making and repair (Mullins 1981; Troy 2003).

Retreat into the autonomy of private homes and gardens distended Australian cities, and the nation's capitalist economy, only amplifying suburban desire for the margins (Davison 1995). The proportion of homes in Australia occupied by owners (outright and mortgaged) at the 2011 Census was 67 per cent (ABS 2012b), among the highest in the world. No less than 86 per cent of these homes are detached houses embedded in private outdoor space, predominantly in low-density cities (ABS 2012b). The sovereignty of private property in land has done much to underpin the ontological conditions of colonial Australian belonging. In the making of this householder democracy, tangible suburban boundaries, written in law and in fence palings, were at one with the imaginative birth of a settler nation-state (Allon 2008). The bounding of the home in Australia through capitalist ownership has simultaneously been an act of private retreat and of public solidarity.

For many living in cities, the roots of settler Australian identity stretched out-back, where European dreams of a rural idyll were refashioned by pioneer figures such as the drover and bushman, who engaged in a retrospectively romantic bare-handed tussle with nature. The declaration of nationhood in 1901 saw a nascent cultural reliance on ecological nativeness in accounts of Australian belonging (Franklin 2006a). This reliance was to grow and later to appropriate Indigenous cultures in an iteration of colonial dispossession (Head 2000). The environmentalist movements that shadowed the second suburban boom in the decades after World War II cultivated romance for ancient wilderness from their suburban heartlands (Davison 2006). Despite rapid population growth, this boom saw the proportion of Australian homes that were privately owned increase rapidly to a peak of 71 per cent in 1966 on the back of mortgage tenure (Troy 2000, 719). This boom also saw the suburban peasantry dismantled, allowing romantic dreams of nature to flourish in private homes and gardens whose role shifted decisively from production to consumption (Gaynor 1999). Yearning for an urban antithesis, for a sanctuary devoid of technology as much as it is home to wild nature, is today widespread in Australian cities (Davison 2008). Yet, at the same time, environmentalist movements have documented the ecological destruction caused by widespread retreat into expansive homes and gardens in vast, sprawling cities during the post-war boom. In the context of fossil fuel-induced climate change, the 'car dependency' of Australian cities has been central to criticism of the unsustainability of suburban lives. Ironically, however, anti-suburban sentiment has long fuelled suburban development in Australia (Davison 2013; Gilbert 1988), and environmentalist antipathy for suburban lives is having complex and perverse effects (Davison 2008, 2011).

Suffice it to say, in the making of Australian modernity, irony is the rule and paradox the mechanism. Central to this paradox, the boundaries of the private home have been part of the effort of modern Australians to live at a safe distance from the profitable but disturbing realities of technological domination and colonial dispossession in an alien land. Crucial to the ability of the material and semiotic boundaries of home ownership to establish this distance has been the lack of critical scrutiny they have attracted. As recently as 1983, Jim Kemeny (1983, 2) observed that Australian home ownership had 'been little studied.' This observation no longer holds (Badcock 2012; Jacobs 2015), although the body of research on home ownership arguably remains modest and partial, bearing out still Kemeny's (1983, 2) claim that 'the very universality' of Australian home ownership protects it from investigation. The neglect of home ownership throughout much of Australia's urban history is paralleled by an equally remarkable neglect of suburban life, which has attracted a good deal of intellectual disdain but little meaningful study until recently (Davison 2006). And while there is some excellent research on the more-than-human elements of Australian homes (Franklin 2006b; Power 2009a, 2009b, 2012), the links between home natures, wider urban natures and home ownership have been little studied. These links are, however, becoming increasingly visible as the dualism between city and nature that has underpinned an Australian search for home in the margins between them has come under wide challenge.

Remaking urban nature in Australia

Physical evidence of a transformation of Australian urban nature over the past three decades is diverse, present in everything from the rise of municipal urban forestry to the renewed biodiversity of suburban creeks; from energy-efficient high-rise apartments to promenading tourists on post-industrial waterfronts; from roof-top bee hives to suburban vistas of photovoltaic panels; from farmer's markets to the labels on supermarket shelves; from the growth of bicycle infrastructure to showrooms of electric cars. This evidence is present also beyond city limits: in droves of urban tourists in chic wilderness resorts; in satellite monitoring of ocean levels; in pallets of organic Peruvian asparagus en route to Australian supermarkets. Conceptual evidence of this transformation is equally pervasive, on display in everything from metropolitan plans to the patter of lifestyle television hosts to school curricula.

The concept of urban nature is now busy in Australian environmental discourses of ecology, eco-efficiency, resilience and adaptation (Lehman and John 2014; Newman *et al.* 2009), albeit often in the guise of derivative concepts such as green infrastructure, ecosystem services and socio-ecological systems (Byrne *et al.* 2014). Much of this interest in urban ecology has been driven by civil society and linked to discourses of community and place-making (Dhakal 2014). Less overtly, new ideas about urban nature are becoming central to economic discourses of liveability, health, security, risk and competitiveness (Hu 2015; Lowe *et al.* 2015; Newton 2012), and to political discourses of equity and justice (Mee *et al.* 2014; Steele *et al.* 2012). These disparate environmental, social, economic and political discourses inform increasingly complex policy agenda for urban sustainability in Australia. These agenda are both cause and consequence of an ongoing transformation of Australian urban nature by processes of consolidation, deindustrialisation and financialisation. Yet, to this point, these agenda have left the practices of home ownership, and the ontological boundaries of private life they underpin, unquestioned.

Urban consolidation in Australia has been driven primarily by a curious combination of environmental concern about suburban form and neoliberal capital accumulation. Implemented via a mix of market opportunism and planned intent, its consequences are contested (Dhakal 2014; Newton 2012). In part, consolidation has been achieved by a decline in the size of suburban land blocks, although the average size of Australian houses has increased, thus ensuring a significant reduction in private garden space (Hall 2010). Consolidation has also been facilitated by the de-industrialisation that is changing the metabolic functions of urban nature and enabling redevelopment. A shift from manufacturing to services promises to underwrite improvements in local air, soil and water quality in cities while imbricating them further in opaque geographies of pollution in 'developing' nations. It has altered the coproduction of urban and nonurban space, with the nation's reliance on resource extraction in remote parts of the continent ever more obscured in cities claiming a new dispensation for nature. The detoxifying and prettifying of urban nature has been a key element in reinventing Australia's cities as honey-traps in intensifying flows of footloose capital. Urban nature is now vital to emerging cultural dynamics of capital accumulation centred on experiential commodities such as liveability, leisure, luxury, exclusivity and creativity (Dovey 2005). These dynamics rely on 'nature' to underwrite the authentic distinctiveness of urban places, while at the same time making these places, and particularly the private homes they contain, readily accessible in global financial regimes.

During the 1980s and 1990s advocacy for urban sustainability in Australia predominantly focused on the resource and waste implications of built form, and particularly on the perceived benefits of urban consolidation (Davison 2006). Despite being centrally concerned with housing form, this advocacy had little, if anything, to say about the power relations of private property and land tenure, and reduced urban nature to ecological inputs and outputs in urban infrastructure. While such advocacy remains influential (Newman *et al.* 2009), the early years of this century have seen the limits of technocratic 'sustainability-as-density' agenda exposed (Quastel *et al.* 2012). Whole-of-life analysis of environmental impacts, rather than analysis focused solely on housing-related resource use, has made clear that income, not spatial location, is the prime correlate for individual 'ecological footprints' in Australia, and that inner-city and inner-suburban areas exert the highest per capita environmental pressure (Dey *et al.* 2007). Growing cultural research interest in home-making practices, behaviours and lifestyles and their relevance to urban sustainability (Lane and Gorman-Murray 2011) is also challenging the determinist assumption that human behaviours can be engineered by planners through urban form.

The neoliberal shift of political discourse from the rights and responsibilities of citizens to the rights and responsibilities of consumers has also foregrounded the relationship between everyday practices and urban sustainability. Given the hegemony of home ownership in Australia, it is not surprising that 'the householder' is an increasing, if uncritical, focus of environmental policy and planning (Fielding *et al.* 2010; Lane and Gorman-Murray 2011; Maller *et al.* 2012). The identity of the householder does not necessarily exclude tenure other than private ownership. Nonetheless, these policies typically assume householders – and to a much lesser extent landlords (Gabriel and Watson 2012) – have capacities and incentives associated with private economic ownership (Fielding *et al.* 2010; Maller *et al.* 2012; Mee *et al.* 2014).

Sustainability-as-density agenda draw heavily on the anti-suburbanism that paradoxically thrives in Australia. Reflecting this paradox, many environmentalists offer scathing critiques of suburban sprawl and capitalist consumerism yet value highly the contact with nature provided by the capitalist practices of suburban home and land ownership: in private gardens and related practices of cultivation, stewardship self-provision and outdoor living; in the companionship of nonhuman family members and the presence of wild nonhumans; in the seasonal recreation and vistas provided by suburban coasts, mountains, waterways, wetlands and bushland; and in the proximity of city limits and the presence of the great beyond (Davison 2008). It is to the importance of urban nature in everyday life that I now turn through an exploration of the practices of home ownership as sources of ontological security in the midst of the ontological anxieties of settlement, anxieties now fuelled by concerns about the sustainability of the modern project of progress.

Securing private nature

It is widely accepted in housing scholarship (Saunders and Williams 1988; Smith 2015) that home ownership functions as a source of ontological security in late-modern societies in which responsibility for self-identity has shifted from collective contexts to self-made individuals (Giddens 1991). This function is inflected, and arguably made more acute, in New World societies such as Australia, in which the settler assertion of possession confronts still-active legacies of dispossession (Crabtree 2013). Anthony Giddens (1991, 36) observes that ontological security exists in a dialectic with the ontological anxiety resulting from the chaos that shadows any social order: 'And this chaos is not just disorganisation but the loss of the sense of the reality of things and of other persons.' In the case of Australia, the ontological security of home ownership has been a crucial countervailing force to anxieties activated by the rapid progress of modernisation in a geography of invasion.

The sovereign home in the margins of Australian cities was a key part of an ambivalent cultural engagement with modernity in settler society, as well as with an unintelligible Indigenous order (Davison 2006, 2008, 2011). The modern displacement of God by reason, of tradition by novelty, of nature by machine, and of particularity by universality was neither complete nor uncontested. Rather, the creation of a public realm founded on secular, dispassionate reason went hand-in-hand with a diurnal rhythm of suburban withdrawal into a private realm founded on sentiment, embodiment, faith and heritage. Suburban homes and gardens have constituted a geography of innocence into which the foot soldiers of modernity retreated from the uncompromising (masculine) project of control, prediction and efficiency on which they were embarked into the (feminine) sanctuary of 'the family', 'the community', 'the garden' and 'the spirit'. While private suburban home ownership has spun the wheels of capital accumulation and underpinned nation-building, it has simultaneously counteracted the reduction of nature and humanity to instrumental means in service to the end of progress (Horkheimer and Adorno 1969). It does so by creating powerfully fortified boundaries around a private realm in which applies moral judgements and affective relations very different from those operating in the public realm. In this way private home ownership facilitated the progress of Australian modernity by incorporating and deactivating within it the disquiet of those who instituted it. Violent excesses in the making of Australian modernity can only be understood in the context of the existential comforts afforded by autonomous homes placed ambivalently, betwixt and between, the worlds of nature and technology.

The ontological security promised by home ownership in Australian cities in the twenty-first century is no less significant than that offered during the previous two centuries. And it is no less paradoxical. It is, however, a dynamic ontological practice, one constituting and constituted by emergent and contingent cultural, material and political realities. At present, attention is understandably being paid to financial expressions of the paradoxical security offered by home ownership (Aalbers 2015; Cook *et al.* 2013; Smith 2015), particularly in neoliberalising societies such as Australia, in which public welfare is weakening (Badcock 2012) and housing affordability declining (Jacobs 2015). Put simply, territorial home ownership (Blomley 2015) is relied upon to produce financial security through private capital accumulation, thereby enabling goods such as leveraged consumption and investment, provision for old-age and intergenerational endowment. Yet home ownership is increasingly founded upon deterritorialised flows of debt and speculation that escalate financial risks to such extent that global capitalism itself quaked in 2008, with ongoing effect.

The embedding of home ownership in financial flows of investment, debt, speculation and accumulation is one aspect of the wider capitalist metabolism of the Urbanocene in which the whole planet is impoverished by an uneven process of urban wealth creation (Gleeson 2015). The planetary economic effects of home ownership deserve scrutiny within urban political ecology. However, the economic function of home ownership in the capitalist metabolism of nature is far from being the only way in which home ownership and planetary futures are co-produced. My aim here has been to draw attention to the neglected role of home ownership in constituting the ontological politics of nature through processes such as colonial-belonging and nation-building.

Accumulating evidence of systemic ecological and geological change stemming from the advance of modernity is a powerful emerging source of ontological anxiety. The effects of climate change – extreme weather events, sea-level rise, species extinction and the rest – are, of course, deservedly rational sources of anxiety, as are related concerns about the exhaustion of freshwater, food and energy resources, the potential for pandemics, the prevalence of environmental toxicity and the irreversibility of ecological unravelling. Many environmentalists explicitly seek to inflame public anxiety about impending catastrophe in the belief that fear will motivate urgent and radical efforts to mitigate these problems. In this context, it needs to be asked whether the practice of ontological security through home ownership counteracts such ecological anxieties. This leads to the more fundamental and deeply troubling question of whether escalating ecological concern escalates the pursuit of private home security in ways that disable political responses to these concerns, causing the collective ecological predicament to further escalate.

In Australia, the lack of transformative government action and democratic mobilisation on many global environmental concerns, and particularly on climate change, is juxtaposed with growing policy and cultural emphasis on the role of householders in cutting their own paths towards sustainable ways of living in their homes. Having its immediate roots in countercultural movements in the 1960s and 1970s (Davison 2011), widespread optimism in the capacity of householders to lead social transformation began to take hold in Australia in the 1990s. This optimism is exemplified by the widespread media exposure given to one inner-city, middle-class family in Sydney pursuing household sustainability (Mobbs 2010). As Pat Troy

(2003) has argued, this new source of ontological security afforded by home ownership arcs back to the history of Australia's suburban peasantry. Long-established practices of animal husbandry, water harvesting, organic gardening, resource re-use and local barter economies were displaced from Australia's cities not long before post-war countercultural movements emerged to denounce suburban materialism.

Household sustainability initiatives founded on the practices of home ownership offer one way in which private encounters with nature are affording many Australians a sense of agency in the face of global ecological anxieties. There is hope to be drawn from this agency if citizens are enabled to gain purchase on the political economic roots of collective problems, although this mode of empowerment is linked to the disempowerment of the growing proportion of Australians excluded from home ownership (Mee *et al.* 2014). The desire of urban households to actively participate in the management, production and conservation of the resources on which they depend has potentially profound implications for public policy, planning and design for urban sustainability, as well as for the activation of participatory civil society.

As this chapter has sought to make plain, such connections between private and public aspiration are far from assured. They need to be the focus of collective political work. Without explicit political intent and action, current interest in sustaining private natures in Australia will continue to serve paradoxical logics of private financial security; in the form, say, of asset appreciation and 'green' brand distinction in real estate markets. They will serve paradoxical logics of private future-proofing; in the form, say, of insurance policies and retreat to higher-ground in coastal cities. They will serve paradoxical logics of private redemption; in the form, say, of caring for injured wildlife at home and donating to private nature conservation organisations. They will serve paradoxical logics of private morality; in the form, say, of seeking ethical consumption choices in the capitalist marketplace. For, as I have argued here, retreat behind the boundaries of a private realm of home premised on political and economic sovereignty is a core feature of Australian modernity, and a constitutive element of the public world from which refuge is understandably being sought.

Note

1 I use the idea of 'urban nature' to broadly encompass inner-urban, suburban and peri-urban places and processes. Rather than implying a division between urban and nonurban natures, I understand them to be coproduced dialectically.

References

Aalbers, M. 2015. The great moderation, the great excess and the global housing crisis. *International Journal of Housing Policy* 15(1): 43–60.
ABS 2012a. *Regional Population Growth, Australia, 2011.* Australian Bureau of Statistics, Canberra.

ABS 2012b. *Basic Community Profile: Australia. 2011 Census of Population and Housing*. Australian Bureau of Statistics, Canberra.

Allon, F. 2008. *Renovation Nation: Our Obsession with Home*. University of New South Wales Press, Sydney.

Angelo, H. and Wachsmuth, D. 2015. Urbanizing urban political ecology: a critique of methodological cityism. *International Journal of Urban and Regional Research* 39(1): 16–27.

Badcock, B. 2012. Beyond home ownership: housing, welfare and society. *Urban Policy and Research* 30: 341–344.

Blomley, N. 2013. Performing property: making the world. *Canadian Journal of Law and Jurisprudence* 26(1): 23–48.

Blomley, N. 2015. The territory of property. *Progress in Human Geography*. DOI: 10.1177/0309132515596380.

Braun, B. 2005. Environmental issues: writing a more-than-human urban geography. *Progress in Human Geography* 29(5): 635–650.

Butlin, N.G. 1964. *Investment in Australian Economic Development 1861–1900*. Cambridge University Press, Cambridge.

Byrne, J., Sipe, N. and Dodson, J. (eds) 2014. *Australian Environmental Planning: Challenges and Future Prospects*. Routledge, Abingdon.

Cook, N., Smith, S.J. and Searle, B.A. 2013. Debted objects: homemaking in an era of mortgage-enabled consumption. *Housing, Theory and Society* 30(3): 293–311.

Crabtree, L. 2013. Decolonising property: exploring ethics, land, and time, through housing interventions in contemporary Australia. *Environment and Planning D: Society and Space* 31: 99–115.

Davison, A. 2001 *Technology and the Contested Meanings of Sustainability*. State University of New York Press, Albany, NY.

Davison, A. 2006. Stuck in a cul-de-sac? Suburban history and urban sustainability in Australia. *Urban Policy and Research* 24: 201–216.

Davison, A. 2008. The trouble with nature: ambivalence in the lives of urban Australian environmentalists. *Geoforum* 39: 1284–1295.

Davison, A. 2011. A domestic twist on the eco-efficiency turn: technology, environmentalism, home. In R. Lane and A. Gorman-Murray (eds) *Material Geographies of Household Sustainability*. Ashgate, London, 35–49.

Davison, G. 1995. Australia: the first suburban nation. *Journal of Urban History* 22: 40–74.

Davison, G. 2013. The suburban idea and its enemies. *Journal of Urban History* 39(5): 829–847.

Dey, C., Berger, C., Foran, B., Foran, M., Joske, R., Lenzen, M. and Wood, R. 2007. Household environmental pressure from consumption: an Australian environmental atlas. In G. Birch (ed.) *Water Wind Art and Debate: How Environmental Concerns Impact on Disciplinary Research*. Sydney University Press, Sydney.

Dhakal, S.P. 2014. Securing the future of urban environmental sustainability initiatives in Australia. *Urban Policy and Research* 32(4): 459–475.

Dovey, K. 2005. *Fluid City: Transforming Melbourne's Urban Waterfront*. University of New South Wales Press, Sydney.

Fielding, K.S., Louis, W.R., Warren, C. and Thompson, A. 2010. *Environmental Sustainability: Understanding the Attitudes and Behaviour of Australian Households*, AHURI Final Report No. 152. Australian Housing and Urban Research Institute, Melbourne.

Franklin, A. 2006a. *Animal Nation: The True Story of Animals and Australia.* University of New South Wales Press, Sydney.

Franklin, A. 2006b. 'Be[a]ware of the dog': a post-humanist approach to housing. *Housing, Theory and* Society 23: 137–156.

Gabriel, N. 2014. Urban political ecology: environmental imaginary, governance, and the non-human. *Geography Compass* 8(1): 38–48.

Gabriel, M. and Watson, P. 2012. Supporting sustainable home improvement in the private rental sector: the view of investors. *Urban Policy and Research* 30(3): 309–325.

Gandy, M. 2012. Queer ecology: nature, sexuality, and heterotopic alliances. *Environment and Planning D: Society and Space* 30: 727–747.

Gaynor, A. 1999. Regulation, resistance and the residential area: the keeping of productive animals in twentieth-century Perth, Western Australia. *Urban Policy and Research* 17(1): 7–16.

Giddens, A. 1991. *Modernity and Self-Identity: Self and Society in the Late Modern Age.* Polity, Cambridge.

Gilbert, A. 1988. The roots of anti-suburbanism in Australia. In S.L. Goldberg and F.B. Smith (eds) *Australian Cultural History.* Cambridge University Press, Cambridge.

Gleeson, B. 2015. *The Urban Condition.* Routledge, Abingdon.

Grove, K. 2009. Rethinking the nature of urban environmental politics: security, subjectivity, and the non-human. *Geoforum* 40: 207–216.

Hall, T. 2010. Goodbye to the backyard? The minimisation of private open space in the Australian outer-suburban estate. *Urban Policy and Research* 28(4): 411–433.

Harvey, D. 1996. *Justice, Nature and the Geography of Difference.* Blackwell, Malden, MA.

Head, L. 2000. *Second Nature: The History and Implications of Australia as Aboriginal Landscape.* Syracuse University Press, Syracuse.

Heynen, N. 2014. Urban political ecology I: the urban century. *Progress in Human Geography* 38(4): 598–604.

Heynen, N. and Perkins, H.A. 2005. Scalar dialectics in green: urban private property and the contradictions of the neoliberalization of nature. *Capitalism, Nature, Socialism* 16(1): 99–113.

Heynen, N., Kaika, M. and Swyngedouw, E. (eds). 2006. *In the Nature of Cities: Urban Political Ecology and the Politics of Urban Metabolism.* Routledge, London and New York.

Horkheimer, M. and Adorno, T. 1969. *Dialectic of Enlightenment*, trans. J. Cumming. Continuum, New York.

Hu, R. 2015. Sustainability and competitiveness in Australian cities. *Sustainability* 7: 1860–1880.

Jacobs, K. 2015. The 'politics' of Australian housing: the role of lobbyists and their influence in shaping policy. *Housing Studies.* DOI:10.1080/02673037.2014.1000833

Keil, R. 2003. Urban political ecology. *Urban Geography*, 24(8): 723–738.

Kemeny, J. 1983. *The Great Australian Nightmare.* Georgian House, Melbourne.

Kirkpatrick, J.B., Davison, A. and Daniels, G.D. 2013. Sinners, scapegoats or fashion victims? Understanding the deaths of trees in the green city. *Geoforum* 48: 165–176.

Lane, R. and Gorman-Murray, A. (eds) 2011. *Material Geographies of Household Sustainability.* Ashgate, London.

Latour, B. 2004. *Politics of Nature: How to Bring the Sciences into Democracy*, trans. C. Porter. Harvard University Press, Cambridge, MA and London.

Lehmann, S. and John, M. 2014. Green urbanism, zero waste and ecological connections. In J. Byrne, N. Sipe and J. Dodson (eds) *Australian Environmental Planning: Challenges and Future Prospects*. Routledge, Abingdon.

Lowe, M., Whitzman, C., Badland, H., Davern, M., Aye, L., Hes, D., Butterworth, I. and Giles-Corti, B. 2015. Planning healthy, liveable and sustainable cities: how can indicators inform policy? *Urban Policy and Research* 33(2): 131–144.

McCarthy, J. 2005. First world political ecology: directions and challenges. *Environment and Planning A* 37: 953–958.

Maller, C., Horne, R. and Dalton, T. 2012. Green renovations: intersections of daily routines, housing aspirations and narratives of environmental sustainability. *Housing, Theory and Society* 29(3): 255–275.

Mee, K., Instone, L., Williams, M., Palmer, J. and Vaughan, N. 2014. Renting over troubled waters: an urban political ecology of rental housing. *Geographical Research* 52(4): 365–376.

Merrill, T.W. 2004. Private property and the politics of environmental protection. *Harvard Journal of Law and Public Policy* 28(1): 69–80.

Meyer, J. 2009. The concept of private property and the limits of the environmental imagination. *Political Theory* 37(1): 99–127.

Mobbs, M. 2010. *Sustainable House*, 2nd edn. Choice Books, Sydney.

Mol, A. 1999. Ontological politics: a word and some questions. In J. Law and J. Hassard (eds) *Actor Network Theory and After*. Blackwell, Oxford.

Mullins, P. 1981. Theoretical perspectives on Australian urbanisation I: material components in the reproduction of Australian labour power. *Australian and New Zealand Journal of Sociology* 17(1): 65–76.

Newman, P., Beatley, T. and Boyer, H. 2009 *Resilient Cities: Responding to Peak Oil and Climate Change*. Island Press, Covelo.

Newton, P.W. 2012. Liveable and sustainable? Socio-technical challenges for twenty-first-century cities. *Journal of Urban Technology* 19: 81–102.

Plumwood, V. 1993. *Feminism and the Mastery of Nature*. Routledge, London.

Power, E. 2009a. Domestic temporalities: nature times in the house-as-home. *Geoforum* 40: 1024–1032.

Power, E. 2009b. Border-processes and homemaking: encounters with possums in suburban Australian homes. *Cultural Geographies* 16: 29–54.

Power, E. 2012. Domestication and the dog: embodying home. *Area* 44(3): 371–378.

Quastel, N., Moos, M. and Lynch, N. 2012. Sustainability-as-density and the return of the social: the case of Vancouver, British Colombia. *Urban Geography* 33: 1055–1084.

Saunders, P. and Williams, P. 1988. The constitution of the home: towards a research agenda. *Housing Studies* 3: 81–93.

Smith, N. 2006. Foreword. In N. Heynen, M. Kaika and E. Swyngedouw (eds) In *The Nature of Cities: Urban Political Ecology and the Politics of Urban Metabolism*. Routledge, London and New York.

Smith, S.J. 2015. Owner occupation: at home in a spatial, financial paradox. *International Journal of Housing Policy* 15(1): 61–83.

Steele, W., Maccallum, D., Byrne, J. and Houston, D. 2012. Planning the climate-just city. *International Planning Studies* 17(1): 67–83.

Troy, P. 2000. Suburbs of acquiescence, suburbs of protest. *Housing Studies* 15(5): 717–738.

Troy, P. 2003. Saving our cities with suburbs. In J. Schultz (ed.) *Dreams of Land: The Griffith Review*. Griffith University/ABC Books, Meadowbank, QLD/Sydney.

Williams, P. 1984. The politics of property: home ownership in Australia. In J. Halligan and C. Paris (eds) *Australian Urban Politics: Critical Perspectives*. Longman Cheshire, Melbourne.

Williams, R. 1973. *The Country and the City*. London, Chatto and Windus.

Wolch, J. 2007. Green urban worlds. *Annals of the Association of American Geographers* 97(2): 373–384.

7   Making nature and money in the East Perth redevelopment

Laurence Troy

Introduction

Over the past 30 years, there has been increasing recognition of the need to address environmental concerns of Australia's cities (McManus 2005). How this is to be achieved remains contentious, yet the rise of sustainability as a concept, for a time at least, offered some hope of an alternate pathway of development that was more aligned with the material realities of ecological systems and processes. One strategy that has been pursued across most Australian cities, and indeed many Western cities, is to renew older industrial precincts to embody emerging notions of the putative sustainable city and emerging economic ambition (Gleeson and Low 2000). The leitmotiv of the sustainability agenda has been to provide higher-density housing in these inner urban 'voids' (Doron 2000) while offering a new 'urban' way of living as an antithesis to supposed unsustainability of suburban living (Davison 2006). The capacity for this ambition to actually reduce the environmental impact of urbanisation has been critiqued by many from a resource-use point of view (Gleeson 2014), yet this can often stand in contrast to the claims being made by governments and developers about the environmental credentials of the redevelopment. This has led some to suggest that these claims have been part of a strategy that is about opening new spaces of accumulation, rather than addressing entrenched unsustainability of cities and economies (Krueger and Gibbs 2007).

Through an exploration of the East Perth redevelopment project in Western Australia, this chapter draws attention to the discursive frameworks through which waterfront housing development in Australia has occurred. Using the framework of urban political ecology, the chapter first conceptualises housing/home as an open-ended socio-natural process. From this vantage point, waterfront housing renewal could evolve in a myriad of different ways, including towards socially and ecologically just ends. However, the chapter argues that three dominant discourses of nature have bound the process of urban transformation and renewal to uneven and exclusionary urban outcomes. These discourses presented nature in East Perth as (1) as an empty space to be cleaned up; (2) a resource whose potential is limited to the creation of elite spaces for wealthy residents; and (3) a mechanism that links public space to private property. Recognising that discourses have material effects (Duncan and Duncan 2001), the chapter argues that these narratives consolidated a set of relationships between private housing and public space and bound these spaces into a process of market exchange. Nature came to embody an exchange value relation establishing a mechanism through which the aesthetic of public space became intrinsically linked to the exchange value of private spaces. In so doing, public spaces came to reflect the desires of a local housing market that was trying to legitimate this new form of housing rather than socially or ethically derived desires in respect of environment and housing and their potential to address many of these concerns.

The research presented in this chapter was undertaken as part of a broader project on the production of nature and sustainability policy in Australian cities. The research focused on two major redevelopment projects that were formative in establishing a new approach to understanding sustainability in urban policy in Australia. The research was based on semi-structured interviews of key people involved in establishing the East Perth Redevelopment Authority (EPRA) and representatives of peak industry organisations, government employees, academics and politicians, all of whom have had direct or indirect roles with the East Perth redevelopment. A total of 13 people were interviewed for this case study and are referred to by their generic role at the time of interviews to ensure some level of anonymity. The remainder of the chapter is split in two broad parts. The first situates this research within an urban political ecology framework which is based around a dialectical conception of nature and society. Within this framework, the production of urban space involves the transformation of nature, where the costs of such transformation are both hidden, and the subject of political concern (Heynen *et al.* 2006a). The second part draws in the East Perth case study to outline three broad discourses of nature used to justify and implement the project.

The nature of change

Nature can be understood in a number of different ways, which have by no means been consistent through time. Harvey (1996, 117), suggests that the 'incoherence of nature as a unitary concept' signals the 'extraordinary amount of human history' contained within words like 'nature', but more importantly reveals a set of values, standards and beliefs that is contained within its meaning. Castree (2005, 8) proposes that there are three conventional definitions or interpretations of nature: first the non-human world; second, 'nature' as the essence of something; and third, 'nature' as an inherent force. Common to Castree's three conventional conceptions is the idea that nature is 'out there' where the nonhuman world is external to what we may call society (Castree and Braun 1998; Smith 1984). It is the material world around us that is waiting to be shaped by human processes, to be dominated through the production process and used as humans see fit. 'External nature is pristine, God-given, autonomous; it is the raw material from which society is built, the frontier which industrial capitalism continually pushes back' (Smith 1984, 11). Castree and Braun (1998, 7) suggest that representations of nature as external underpin 'the "manage nature" discourses of technocentrics in government, business and the like'. Simultaneously, and somewhat in contradiction to 'external nature', the second two definitions above conceive nature as 'universal'. This includes human beings as biological entities, mere blips in the evolutionary movements of the earth, with social processes becoming collapsed into 'laws of nature' and therefore immutable. This conception of nature positions humans themselves as the problem and removing humans becomes the only way to 'save' universal nature from destruction.

Both external and universal natures maintain a dualism between nature and society, positioning them as distinct ontological entities that may exist independently of one another. The simplest of these dichotomies is embodied in the distinction between city and country, society on the one hand and nature on the other. Cities are an embodiment of everything social, while country, such as national parks, embodies everything that is natural. These authors maintain that these conventional dualist ontologies are problematic and insist on conceiving nature in dialectic with human existence (Heynen *et al.* 2006a) so that it is impossible to conceive of a material or 'natural' world independent of human history and human existence and where human existence cannot be understood outside of its 'natural' context. This is not to deny ecological processes independent of human existence; rather, it is to insist that society and all this encompasses is deeply implicated in nature. This has shaped both ways of understanding nature and its material transformations.

The point about opening the discussion on conceptions of nature is twofold. First, it is to propose a re-evaluation of our understanding of cities as purely social spaces and therefore devoid of real nature. Cities and urban environments are what Swyngedouw (1999) calls 'socionatures', that is, they are 'hybrid, part social, part natural – yet deeply historical and thus produced – objects/subjects are intermediaries that embody and express nature *and* society and weave networks of infinite liminal spaces' (Swyngedouw 2006, 25; original emphasis).

Second, a key concept running through the idea of nature is that human and nonhuman elements are implicated in both producing and being produced by the material world (Heynen *et al.* 2006b; Smith 1984). In this way, it is practically and ontologically impossible to separate conditions in nature, and therefore problems in nature, from the very process and relations that govern the organisation and regulation of human existence (Gunder and Hillier 2009; Swyngedouw 2007). This means that processes of socio-economic organisation are directly implicated in producing nature, the material world; as such divorcing discussion of ecological problems from debate on socio-economic organisation will only serve to externalise many social foundations of environmental problems.

This of course is not to deny the many technical and administrative changes that have had positive implications for reducing the consumption or destruction of material resources. Rather, the suggestion here is to expand the field of analysis and understanding of problems faced in urban areas and how they can be addressed to also include different forms of socio-economic organisation. Importantly, this concept of socio-nature allows us to directly implicate the very material spaces that are commonly conceived as epitomising all things 'un-natural', such as urban areas. Harvey's famous edict on New York, proclaiming it to be, at its core, natural (Harvey 1993), forces a re-evaluation of urban spaces and how they are entwined with the metabolic process and circulation in nature, and the political and economic context through which it comes into existence. It also implicates those things that are often thought of as unnatural, such as housing, into discussions about socio-natural (urban) futures. It speaks to the very title of this book by 'unbounding' house and home from only social conceptions of the city and productions of bounded space encapsulated in the idea of house, and inserting them in wider narratives of urban ecological change.

The urban political ecology narrative, however, is more than reorienting society–nature ontologies, it is also about providing a framework for understanding specific socio-natural relations under capitalism. It is about understanding how capitalism brings 'all manner of natural environments and concrete labour processes upon them together in an abstract framework of market exchange' (Castree 1995, 20). One of the key ways this has been achieved is through the privatisation and commodification of nature (Heynen *et al.* 2007). For Castree (2003, 277), it is not only to understand that something is a commodity, it is also to understand how it, in this case nature, has been commodified and the specific properties it takes on when commodified. Understanding that the commodity status of an object is not intrinsic, but rather assigned, leads us to question the social, political and economic context in which this is occurring.

This framework opens up two important lines of inquiry about the production of new urban natures. First it brings into question the increasing focus on the physical design-led interventions into managing urban change that tends to dominate planning discourse in Australia. The preoccupation with density as a physical outcome within a sustainability discourse is one example of policy that rarely focuses on reduced consumption and growth dynamics with a political and economic context (see Gleeson 2014; Gleeson and Low 2000). Swyngedouw (2007, 2009) describes the process as 'post political', in which debate on alternative urban futures, particularly around issues of sustainability, are effectively limited to the technological or administrative reorganisation of space, and precludes questioning, let alone changing, the socio-economic foundations that underpin urban processes. In this context, excluding politics from discussion on urban futures in effect confines discourse and disagreement to technocratic or design outcomes.

Second, it opens question about precisely 'whose nature and whose culture' (Katz 1998) are being attached to understandings of value within this framework. In the urban context, where natures are being reconceptualised and remade in a very material sense, questioning 'whose nature' should alert us to the politics implicated in the production of these new spaces. Duncan and Duncan (2001) have noted the important way landscape is implicated in this process as a positional good which brings to the fore the complex politics of landscape, identity and aesthetic. They argue that sense of place and place attachment combine with 'political-economic flows and processes that are central to place production' (Duncan and Duncan 2003, 4). In the example of Bedford, near New York City, they examine how aesthetic appreciation of wilderness and of landscape was acting as a mechanism for social exclusion and perpetuation of elite identities. The discourses of landscape and wilderness were integral to contesting the planning ambitions of some, and ultimately excluding others in efforts to maintain this idealised vision of landscape.

How, then, do discourses of nature work in relation to waterfront housing redevelopment in Perth? What work do they do in orienting housing/home to particular political-economic outcomes? It is at this point that I turn to the East Perth redevelopment project to explore the relationship between politico-economic flows, discourses of nature and processes of place production in strategies of urban waterfront development. The story opens with a short historical narrative of East Perth and its inhabitants. The reason for this introduction is because the landscape and its shifting identities have formed a key part of the claims being asserted through the redevelopment process and contestations over the future of the area.

East Perth, aesthetics and nature

The East Perth Redevelopment Area covers approximately 140 ha and forms part of the larger East Perth suburb (see Figure 7.1). Prior to European invasion, Perth and its surrounding areas were characterised by a series of wetlands and swamps which were interconnected and drained into the Swan River via a stream initially called 'Clause Brook' and subsequently renamed Claisebrook. The Claisebrook itself was one of the few continuous freshwater supplies in the area and was of both practical and mythological significance for local Indigenous people (Meagher and Ride 1980; O'Neill 2001).

European invasion saw the area transformed into one of the main industrial precincts for Perth, providing necessary products to the small isolated colony (Byrne and Houston 2005; O'Neill 2001; Thomas 1978). From the 1950s, economic and social change began affecting the distribution of land use activity in Perth, brought about by a range of economic and technological factors. By the end of the 1980s, the area, although seemingly in disrepair, provided significant accommodation for the urban poor of the region, most notably the local Indigenous population (Byrne and Houston 2005; Hillyer 2001).

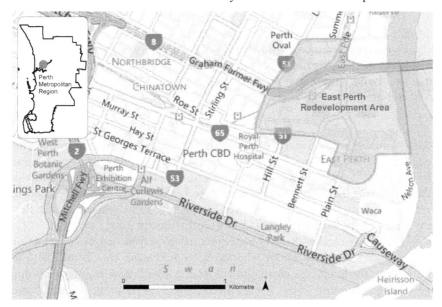

Figure 7.1 Location of East Perth.
Source: basemap retrieved September 2015 from www.bing.com.

In 1985 the state government established the East Perth Landuse and Landscape Committee to investigate redevelopment options for this area; by 1990 major plans to redevelop the area into a new high-tech, mixed-use precinct were proposed. The plans were both intended to provide a form of economic stimulus, representative of a shift in economic focus of the state economy more broadly, and to offer an alternative urban development program in which urban consolidation was being promoted as the key sustainability agenda. In 1991 the EPRA was established through an Act of Parliament that gave the new organisation the legal and financial capacity to make, consent to and deliver plans. Throughout the planning and development process there were a number of discourses that emerged, many of which were constructed around both positive and negative aesthetics of nature and landscape and what it comprised. These discourses were fundamental in defining the scope and character of the final development, and are considered next.

Cleaning the void: nature as something to be cleaned

Throughout the planning process, East Perth was routinely constructed by officials, consultants and developers as a space that was empty and abandoned. This void was important in positioning East Perth as open or ripe for development, a featureless plain, ready to receive the utopian vision of new urban village.

Thing was because it was such an abandoned area of Perth, the city had no view on it, other than it was Perth's backyard and it had become completely redundant and they didn't really know what to do with it. I mean they had a small resident population up here on the hill, a very small one. They had, this had all basically become derelict and very low end uses.

(Consultant)

Although there was a lived presence (if often informal) of the Indigenous population, it was as if East Perth – 'completely redundant' – was being recast as *terra nullius*. This was precisely at the time that the High Court of Australia was handing down what is now known as the Mabo decision to refute the notion that Australia was 'empty' in 1788 when the first fleet invaded Port Jackson in Sydney (Byrne and Houston 2005).

Gil Doron suggests there is a politics being played out in constructing these spaces as 'dead zones' that acts to legitimise the 'brutal act' of bulldozing the landscape clean and colonising it with new architectures and new imagery (Doron 2000, 248). Indeed there were a number of significant buildings that were demolished to make way for the development. In recounting a story about one of the churches in the area, one of the key participants suggested:

the day before the Heritage Act came in, we knocked down one of the last churches in East Perth because it was in the way. We knew if we had the Heritage Act come in then we would have to go through that, and probably have to retain it, so we brought in the bulldozers.

(Government agency)

Doron points out that the same economic, social and political processes that were trying to regenerate, revitalise and re-activate these dead zones were the ones that created them in the first place. Demolishing Bennett House, for example, 'erases evidence of the crime' (Doron 2000, 252) that was committed against the Indigenous people of Western Australia over the previous 150 years. The homelessness of Aboriginal people in East Perth, their inadequate shelters, the racist policies that forced them to the margins in the first place, can be explained by the brutal history of British colonialism and its push to extend the economic reach of the British Empire. The same economic colonialism which saw British capital invade new spaces, and redefine others, underpins the system of redevelopment that pushed Aboriginal people out of East Perth to the new margins.

In practical terms there was of course something in East Perth, and conceptualising the place as empty was about emphasising that there was nothing in East Perth of any value. East Perth was not only being conceptualised as 'empty', but as old and dirty as opposed to the future that was new and clean, with distinctly positive and negative connotations attached to each.

The language referencing past incarnations of East Perth were clearly trying to establish a discourse that cast the place as dirty, derelict, obsolete, polluted, infested with vermin and all manner of things negative.

> That organisation has taken a major area of docks and warehouses which served no real purpose – other than as a haven for vermin.
> (Western Australia Parliament 1991, 1380)

> But never the less at the time it was just the realisation that the Swan River was a polluted river, waterway, heavily polluted waterway and was at risk of survival if some actions weren't taken.
> (Property industry peak body)

> I mean this project enabled us to actually clean up what was, the accumulated detritus of over a century of treating that area as a backyard, so from an environmental perspective I think that was important.
> (Former politician 1)

There were, of course, considerable pollution problems that needed to be addressed due to the accumulation of heavy metals and polyaromatic hydrocarbons from the former gas works, power station and other industries. However, in making the case for change, physical degradation of the landscape was presented as the only issue. These dystopic images focused on what were aesthetic perceptions of nature – dirty, old, worthless – politically confined debate about the development to the physical remediation of these spaces.

The EPRA used these discourses to their advantage by producing plans that were not aesthetically positive to convince government and gain political support to advance a particular model of development or particular vision of what East Perth should look like. A former employee of EPRA recounted a story of some of the early plan development and how they leveraged aesthetics in getting the outcome they desired: 'We did lots and lots of plans, ideas, you know we sent up red herrings of you know, the government didn't want to proceed with inlet, we showed them how bad it could look, you know a lot of stuff like that' (government agency).

Due to the expense of actually constructing an inlet, there was apprehension on the part of the state government and cabinet to provide funds to pursue this option. The dystopic visions or 'red herrings' were being produced to show a landscape of poor quality if the inlet was not built, so the government was left with no choice but to provide funding to build the inlet that EPRA desired. The 'red herrings' were in effect another form of dystopian image of a dysfunctional landscape. This dystopian–utopian dichotomy became an important discourse throughout the development, and an effective lever to advocate for change. Importantly, these dichotomous discourses of nature – as absent or present, dirty or clean, dysfunctional or functional – confined debate on possible futures to the ephemeral physical design elements of the development and precluded the ability to question for whom and for what development was being done for.

Nature as elite spaces for wealthy residents

The images that were created and associated with the new clean future of East Perth focused heavily on the form of public realm. The Swan River was the most important aspect of this image creation. There is no doubt that there were considerable pollution issues in this location and that redevelopment of any kind would necessarily involve a clean-up of this area; however, much of the language surrounding the proposals focused on the image of change and creating something that was of maximum aesthetic appeal.

> The East Perth gasworks site was the most contaminated site in Western Australia. Huge issues about what that would do to your marketing, it had contaminated some 6 ha of the Swan River with poly aromatic hydrocarbons.... East Perth as an image was the real east side of town. So one of the issues was, you had to change people's perceptions to say East Perth wasn't what you thought it was.... So the first thing we put in was signs. You know, like Jesus is coming, the developments here. The East Perth marketing image and then we started to clean up things, started to do work.
>
> (Government agency)

This new image was not just about aesthetic quality, it was about creating a particular identity. So when one interviewee referred to East Perth as 'the real east side of town' the change desired was to shift the identity of the area from one associated with Aboriginality, homelessness and the working class. A former politician of that period, reflecting on the motivations of the project, suggested that it was partly being driven by the desire to push these people out:

> the Perth City Council was pushing for redevelopment of East Perth because ... there was an aboriginal component in there and there were quite a few people who were seen by the City Council as socially undesirable including the aboriginal population and they wanted to push them out, basically.
>
> (Former politician 2)

Richard Lewis, who went on to become the Planning Minister in charge of EPRA during its formative years, expressed the view at the time the EPRA Act was being introduced that this project should not address the social needs of the people in the area, and indeed that these groups should be excluded:

The authority should not be used as a vehicle to placate the social injustices perceived to exist in the community, particularly in the area of housing ... if too much Homeswest housing is erected in East Perth, the viability of the East Perth Redevelopment scheme could be threatened.

(Western Australia Parliament 1991, 1365)

This view was supported within EPRA at the time, with one of the key actors suggesting that the reluctance to incorporate public housing in the development was deliberate, so as not to undermine the commercial aspirations of the project.

It was about redevelopment. One of the big issues of course was the fact that the department of housing had land there and from our point of view we wanted to, although we had later on, you know, inclusive housing policies, we didn't want to see it in the first years, become, you know, have one of the flagships as public housing, this had to be driven by the private sector.

(Government agency)

This new development offered up the perfect opportunity to destabilise links to the past and script a new future that was clean, aesthetically appealing, and devoid of the 'undesirable' elements that threatened its commercial success. As suggested in the previous section, the way to do that was to literally wipe the landscape clean, re-image its physical appearance and in the process its social identity.

Ultimately, EPRA was successful at realising these commercial aspirations, the profitability aspirations, and created a housing sub-market that was enormously successful in driving up property values. The increase in value of property in East Perth and surrounding locations exceeded the change experienced elsewhere in the Perth region, and was reported over a period of years in *The West Australian* newspaper as setting new property sale or price records.

The rapidly growing East Perth redevelopment drove house median prices in the inner city area up a healthy 8 per cent in the September quarter, the best result of any Perth Suburb in a subdued market.

(*The West Australian*, 26 October 1996, 1)

East Perth village tops property rise
Perth's House prices are finally on the move, with East Perth outstripping the best of them with a 17.5 per cent increase in prices over the past year, according to the Real Estate Institute of WA.

(*The West Australian*, 30 July 1997, 11)

East Perth Apartments Project 'sets Record Price'.

(*The West Australian*, 7 May 1997, 67)

East Perth Block Prices Soar 40pc
The value of riverside blocks in East Perth have jumped 40 per cent in 18 months, based on sales at the weekend.

(The West Australian, 20 May 1998, 60)

It would appear that through the re-branding the area was delivering this transformation in physical terms. In other words, the physical spaces of East Perth, as an aesthetic experience, were providing the basis for renewed accumulation through the local real estate market.

Far from being problematic, particularly in relation to early affordable housing objectives, the rise in property value, not just in the area of the redevelopment, but the surrounding zone, was seen as fundamental to the success of East Perth.

> The project also had an enormous impact on the value of East Perth, some of the analyses that were done about the time said you know the property values had gone up 25%. The ripple effect of East Perth was fundamental.
>
> (Government agency)

> We knew it became successful when Janet Holmes à Court bought into there, you know bought, got involved, set up the gallery...we had Lamonts set up a restaurant, you had the Chinese consulate moved in.
>
> (Government agency)

Janet Holmes à Court, for a time being the wealthiest woman in Australia, locating in the area was the endorsement by the wealthy elite that EPRA desired. The physical transformation of the once 'derelict' and 'infested rubbish heap' into a leading light in new urbanism and sustainable development (Australian Council of New Urbanism 2006) was also leading the way in a resurgence of property value in Perth that was clearly 'good for business'.

EPRA was aiming to sell a new neighbourhood identity to attract an urban elite. What distinguished the urban redevelopment process from other examples of renewal was the selling of this new, utopian, clean urban nature. This was designed to attract a particular type of propertied resident. In this context, landscape became commodified through a complex relationship between the production of new 'clean' natures and the formation of scripted 'urban' identities: 'Our vision was centred around an idea of an urban village, which were, sort of, a bit like new urbanism today is' (government agency).

Landscape becomes transformed into a kind of 'cultural capital' which communicates particular social identities, particular ways of living and the exclusion of other 'undesirable' identities. Importantly, it is through this process that the socio-political and economic practices that enact exclusionary practices become stabilised (Duncan and Duncan 2001, 390).

Nature: linking public space to private property

In order to sell developable land, there was a strong focus on all the public areas of the redevelopment, including roads, paving, lighting and landscaping. There was a clear view from the beginning that this development needed to achieve an aesthetic standard above developments that had preceded it around Perth.

> what we are trying to achieve by this development; that is, to provide a new jewel in the crown of Perth and to integrate that jewel into the overall crown.
>
> (Western Australia Parliament 1991, 1381)

> There were lots of debates about, you know, the City of Perth would say for the East Perth, you can have lots, you can have any paving stone you like, as long as it's grey. So you know introducing them to urban stone was pretty revolutionary, the fight on that because the city would then have to maintain that, so the argument then, the City gets the rate revenue, gets to maintain the parks, gets to maintain this, which was to them in some ways gold plated, much above the standards of many other parts of the city.
>
> (Government agency)

The colour of paving stone, 'gold plated' or even 'silver plated' urban design and landscaping does not warrant criticism in itself; achieving high-quality urban space is probably a commendable aspiration. However, the rationale for choosing what was to be included in the public spaces of the development and the form and function of the river inlet was articulated through the desire to increase the marketability of the redevelopment. The quality of 'urban' stone paving, expensive lighting and railing systems were all chosen as a way to demonstrate difference between other areas of Perth, and weren't necessarily a reflection of function, durability and ease of maintenance, which could easily be argued as critical to sustainability of urban infrastructure. The point is that this standard was being driven by aspirations of market differentiation to ensure that it become a profitable development. In so doing, creating this unique and therefore potentially more profitable outcome, established the condition through which exclusive, or in other words exclusionary, productions of space are created.

In a similar way, there was a major focus on the river and the creation of a new inlet due to the property premiums associated with waterfront development. In a state that has ample sun, heat and waterways that have been relatively unaffected by industry (with the exception of East Perth), the importance of water edge locations culturally and commercially cannot be understated. This was well recognised by planners, who saw the creation of the Claisebrook inlet as a 'golden' opportunity to bring property premiums back into the development.

The inlet in most cases was decided because of two things – one is there is an original water body through there, you know, the Claisebrook itself. But it was how do you bring value back into the site and how do you extend the value you get from waterfront back into here other than bring the water in, then you get all that value, so that's a logical thing.

(Consultant)

Much of that, you know if you look at the old maps of Perth, it's a swamp, and this was part of that, so you couldn't build on it without a lot of penalty and so you may as well turn it into an asset which is why we ripped out the thing.

(Government agency)

Increasing the amount of land directly adjacent to the water effectively increased the return on this land. While there was a strong, acknowledged and open emphasis on this logic, much of the publicly stated desire to change the form of the Claisebrook drain was articulated through the desire to return the *Claise Brook* to its original 'unaltered', pre-colonial form, reminiscent of what Smith (1984) describes as its first nature:

but they have done a superb job of retaining that and indeed even the Claisebrook itself, the inlet itself, if you go back to the old maps in 1840s there was that inlet there so you know there has been some things that have been done well.

(Former politician)

The 'natural' form of the Claisebrook was more like a swamp or wetland that drained into the Swan River through the Claisebrook, a small ephemeral stream, which is significantly different to the large open inlet that exists today (see Figure 7.2). Irrespective of whether the final form of the Claisebrook is true to the pre-European past, the image of 'pre-nature' and finally the constructed nature itself were providing the basis upon which real estate values could be pushed upwards. The image of a reinterpreted first nature itself was providing the basis for an accumulation strategy articulated through environmental or green aspirations. As suggested in the quote, the river was fundamental to this 'success', so much so that EPRA, now called the Metropolitan Redevelopment Authority (MRA), are seeking to emulate the same inlet design in a zone immediately to the south of the Claisebrook area, in a redevelopment called 'Riverside', the final piece in Perth's eastern gateway.

Castree (2003) suggests that the commodification of nature (or anything for that matter) is only part of the point when it comes to urbanisation. It is also to understand how something has been commodified and the particular relations things take on when they become commodified. This distinction is relevant in East Perth in two ways. First, identities being deliberately inscribed through the construction of new landscapes were critical to its real estate

Figure 7.2 Current-day Claisebrook Inlet.
Source: the author.

success, and arguably this formed the foundation of the relations underpinning the exchange value of the location. Second, through reversing the relation between public and private spaces and constructing landscape not as something to participate in, but something to be observed (Foucault 1977), the scope of the public space was diminished. By positing the public realm in relation to privatised property arrangements, the 'principle elements are no longer the community and public life, but, on the one hand, private individuals and, on the other, the state, relations can be regulated only in a form that is the exact reverse of the spectacle' (Foucault 1977, 216). That is to say the public areas of East Perth, the landscape of East Perth, was not constructed in response to a need to participate in public life in public space, but to procure for the 'individual, the instantaneous view of a great multitude' (Foucault 1977, 216). Participation in public life through the public realm ceases to become the defining feature, but rather 'views' of landscape or public areas from private areas. Views of the inlet, views of the river, views of open space and views of landscape become enrolled in the system of exchange and in a sense become exchangeable themselves (Troy 2014). Views of public space are consumed from the newly constructed private spaces.

These publicly visible spaces were then the principle focus of planning in East Perth. Street paving, median strips, parks, round-a-bouts, the river foreshore and the building facades visible from a publicly accessible position were the object and the subject of planning for East Perth. Plans, design guidelines and the like were all focused on the publicly visible aspects of the development and were prescriptive in terms of what would happen, and what could happen in the case of the actual buildings. Much like architectural plans for a building produces a static vision of built form and dictates what will be constructed, the landscape and design guidelines for the entire development prescribe a static vision of nature or landscape throughout the development – visions which became intrinsically linked to the marketing and property valuation of the area. Urban nature was being produced as a representation of exchange value, reflecting what is termed the 'qualitative change in the relations of production (Smith 1984, 77). There is a certain requirement in the exchange process that the objects of exchange become known, quantifiable and therefore static, and by linking the nature in the public areas to property exchange it, too, in part, must remain static.

In linking this to wider concepts about sustainability in which relations with nature must change or adapt to reflect the environmental imperatives we are faced with, this static condition becomes problematic. Through the production of static visions of urban nature, it undermines the ability to embody a capacity for change to reflect changing social aspirations or environmental condition. To change the physical composition or layout, and perhaps use of public spaces to reflect the need for change – for example, to facilitate localised food, water or energy production or changing demographic requirements – would have the effect of reshaping the commercial relations embodied in the production of these spaces. In East Perth, the relationship of the Claisebrook inlet to the remainder of the development has been linked to valuation of properties with views of the inlet. To change the aesthetics of the location in favour of other uses that may block views or change the aesthetic, for example, may undermine the perceived value of these properties. Adaption would necessarily undermine the socio-economic or socio-ecological relations that have been produced within this landscape. Of course, not all changes would necessarily undermine this socio-economic relation, but the point is that any change would be constrained to one that at the very least maintains the intrinsic link to property value appreciation.

Moreover, through tying the urban landscape to the urban land market, it establishes a mechanism through which ownership can be asserted over these public spaces. Changing form and function of public spaces, or even questioning or challenging perceptions of what the public spaces should be, like blocking someone's view, for example, then has the potential to generate a particularly reactionary form of politics with protection of values and identities tied to the public realm on the basis of property value. It is not that property owners shouldn't have a right to participate in articulating demands over public spaces, but rather the problem is 'owning' property becomes the defining mechanism that establishes the right to make a claim over the urban landscape. In the case of East Perth, where buying property also includes buying into particular identities constructed through landscape, ownership then becomes asserted not just over one's 'backyard', but encompasses the whole gamut of 'public' space and the exclusive identities it now represents. If, however, the fundamental relation between private and public spaces were different, i.e. the protection of, or production of, public spaces as a reflection of public interest and public need in all its diversity, then there is greater potential for a real politics about the future of urban space that may include land owners and other non-owning users of the land alike.

Conclusion

East Perth from the outset was presented as an opportunity to re-design and re-shape physical space. The original name for the working group, East Perth Landuse and Landscape Committee, speaks volumes about how this project was understood, and how its contribution to the Perth area was conceived. While there was an early opportunity and even intention to introduce significant proportions of public housing, before the first bulldozer even arrived, the project was very much focused around 'cleaning' the accumulated physical and social identities of the area. So now when one buys into East Perth, you are not only purchasing a house or an apartment, you are purchasing an 'inner urban' identity, a lifestyle that has been neatly packaged and intrinsically tied to the aesthetics of the public spaces of East Perth. As one interview respondent suggested, it was about 'changing people's perceptions' of what East Perth was, re-attaching it to the central Perth identity, business oriented and 'urban'. These identities being constructed through discourse, and through the material production of East Perth, underpinned the strategy for accumulation. For the right price, East Perth not only offered beautiful homes, but the moral status of living in a sustainable urban village.

Nature as something to be cleaned, nature as an elite space and nature that links private and public spaces in an abstract context of market exchange were all enlisted to produce an urban renewal housing project that advanced financial accumulation and private homeownership. In this setting profit and privilege, not redistribution and inclusion, became the dominant relations underpinning the production of the new housing in East Perth. These discourses of nature, and ultimately the production of new natures, became central to delivering an exclusionary city through linking it in a practical way directly to the property market for new housing. The case thus highlights the way that urban development and planning processes are not only about producing new housing stock, but also the creation of new urban identities. Exclusion here does not mean that particular peoples cannot access public spaces as one might expect in gated spaces, but rather identities constructed through new public landscapes exclude different identities accessing the housing market in the area. In producing a politics of exclusion through the landscape, the public landscape becomes a place of exclusion. Exclusion becomes more than just physical; it is symbolic, political and economic.

References

Australian Council of New Urbanism. 2006. *Australian New Urbanism : A Guide to Projects*, 2nd edn. Australian Council of New Urbanism, Melbourne.

Byrne, J. and Houston, D. 2005. Ghosts in the city: redevelopment, race and urban memory in East Perth. In D. Cryle and J. Hillier (eds) *Consent and Consensus: Politics, Media and Governance in Twentieth Century Australia*. API Network, Perth.

Castree, N. 1995. The nature of produced nature: materiality and knowledge construction in Marxism. *Antipode* 27(1): 12–48.

Castree, N. 2003. Commodifying what nature? *Progress in Human Geography* 27(3): 273.

Castree, N. 2005. *Nature*. Routledge, Oxford.

Castree, N. and Braun, B. 1998. The construction of nature and the nature of construction: analytical and political tools for building survivable futures. In B. Braun and N. Castree (eds) *Remaking Reality: Nature at the Millennium*. Routledge, London.

Davison, A. 2006. Stuck in a cul-de-sac? Suburban history and urban sustainability in Australia. *Urban Policy and Research* 24(2): 201–216.

Doron, G.M. 2000. The dead zone and the architecture of transgression. *City* 4(2): 247–263.

Duncan, J.S. and Duncan, N. 2001. The aestheticization of the politics of landscape preservation. *Annals of the Association of American Geographers* 91(2): 387–409.

Duncan, J.S. and Duncan, N. 2003. *Landscapes of Privilege: The Politics of the Aesthetics in an American Suburb*. Routledge, New York.

Foucault, M. 1977. *Discipline and Punish: The Birth of the Prison*. Penguin, London.

Gleeson, B. 2014. *The Urban Condition*. Routledge, New York.

Gleeson, B. and Low, N. 2000. Revaluing planning: rolling back neo-liberalism in Australia. *Progress in Planning* 53(2): 83–164.

Gunder, M. and Hillier, J. 2009. *Planning in Ten Words or Less: A Lacanian Entanglement with Spatial Planning*. Ashgate, Farnham.

Harvey, D. 1993. The nature of environment: the dialectics of social and environmental change. *The Socialist Register* 29: 1–51.

Harvey, D. 1996. *Justice, Nature and the Geography of Difference*. Blackwell, Oxford.

Heynen, N., Kaika, M. and Swyngedouw, E. 2006a. *In the Nature of Cities: Urban Political Ecology and the Politics of Urban Metabolism*. Routledge, New York.

Heynen, N., Kaika, M. and Swyngedouw, E. 2006b. Urban political ecology: politicizing the production of urban natures. In N. Heynen, M. Kaika and E. Swyngedouw (eds), *In the Nature of Cities: Urban Political Ecology and the Politics of Urban Metabolism*. Routledge, New York.

Heynen, N., McCarthy, J., Prudham, S. and Robbins, P. 2007. *Neoliberal Environments: False Promises and Unnatural Consequences*. Routledge, New York.

Hillyer, V. 2001. Bennett house: Aboriginal heritage as real estate in East Perth. *Urban Policy and Research* 19(2): 147–182.

Katz, C. 1998. Whose nature, whose culture? Private production of space and the 'preservation' of nature. In B. Braun and N. Castree (eds) *Remaking Reality: Nature at the Millennium*. London: Routledge.

Krueger, R. and Gibbs, D. 2007. *The Sustainable Development Paradox: Urban Political Economy in the United States and Europe*. Guilford Press, New York.

McManus, P. 2005. *Vortex Cities to Sustainable Cities: Australia's Urban Challenge*. UNSW Press, Sydney.

Meagher, S.J. and Ride, W.D.L. 1980. Use of natural resources by the Aborigines of south-western Australia. In R.M. Berndt and C.H. Berndt (eds) *Aborigines of the West: Their Past and Their Present*, 2nd edn. University of Western Australia Press, Perth.

O'Neill, G. 2001. Claisebrook, East Perth. *Early Days: Journal of the Royal Western Australian Historical Society* 12(1): 60–72.

Smith, N. 1984. *Uneven Development: Nature, Capital, and the Production of Space*. Blackwell, New York.

Swyngedouw, E. 1999. Modernity and hybridity: nature, Regeneracionismo, and the production of the Spanish waterscape, 1890–1930. *Annals of the Association of American Geographers* 89(3): 443–465.

Swyngedouw, E. 2006. Metabolic urbanization: the making of cyborg cities. In N. Heynen, M. Kaika and E. Swyngedouw (eds) *In the Nature of Cities: Urban Political Ecology and the Politics of Urban Metabolism*. Routledge, New York.

Swyngedouw, E. 2007. Impossible 'sustainability' and the postpolitical condition. In R. Krueger and D. Gibbs (eds) *The Sustainable Development Paradox: Urban Political Economy in the United States and Europe*. The Guilford Press, New York.

Swyngedouw, E. 2009. The Antinomies of the postpolitical city: in search of a democratic politics of environmental production. *International Journal of Urban and Regional Research* 33(3): 601–620.

Thomas, M. 1978. East Perth 1884–1904: a suburban society. In J.W. McCarty and C.B. Schedvin (eds) *Australian Capital Cities: Historical Essays*. Sydney University Press, Sydney.

Troy, L. 2014. (Re)producing nature in Pyrmont and Ultimo. *Geographical Research* 52(4): 387–399.

Western Australia Parliament. 1991. *Parliamentary Debates (Hansard): Legislative Council and Legislative Assembly*. Perth: Government Printer.

8 Displacement as method
Seasteading, tiny houses and 'Freemen on the Land'

Lorenzo Veracini

This chapter is part of a wider exploration of the global settler colonial present (Veracini 2015). It juxtaposes three ostensibly disconnected political sensibilities because they exemplify a settler colonial 'common sense' (Rifkin 2013) and because they are similar responses to crisis and the expectation of crisis. Many have come to believe that what Australian-based theorist Tony Fry (2011) has called the coming 'Age of Unsettlement' is actually coming fast ('Gods be good'). This is a new dispensation, and yet, like settler colonial sensitivities in other eras, these movements advocate pre-emptive displacement as a solution to the prospect of growing social tension and upheaval. Thus, this chapter contextualises these movements with reference to a long-lasting political tradition, what Gabi Piterberg and I have called the 'world turned inside out' in homage to Christopher Hill's metaphor of revolution as a 'world turned upside down' (Hill 1972; Livingston 2010; Piterberg and Veracini 2015; Veracini 2012). In our analysis we interpret the world turned inside out as a political consciousness that abhors revolution but, rather than countering revolution once it has occurred and contesting the same geography, pre-empts it by means of displacement. If the world turned upside down aims to replace one political order with another, the world turned inside out aims to collectively displace to a political order that is other.

The seasteaders, the dwellers of tiny houses – the 'tiny house people' – and the Freemen on the Land do not share much: not a specific set of ideas, not an idiom and certainly not personnel or politics. These three social movements, however, all envisage impending catastrophic scenarios and similarly believe that displacing somewhere else constitutes the only appropriate response to the forthcoming crisis (indeed, they rely on the expectation of crisis to the point that they act, like other 'preppers', as if the crisis was already here).[1] They crucially rely on the internet to propagate themselves but they also rely on ideas and sensitivities that are available in a shared storehouse of ideas that constitutes the settler colonial present. Their ideas need little explanation or proselytising; they resonate widely and are easily mobilised. These movements also similarly focus on the need to shift modes of inhabitance by adopting displacement as a method of political change. In crucial ways they are contemporary reflections of a desire to unbound housing from spatial fixity.

While originating in settler societies elsewhere, these movements are relevant to Australia. They have a growing number of adepts and supporters there, and there is growing mainstream interest in their activities and proposed solutions (see, for example, Barnes 2015; Duke 2015; Glazov 2014). The unsettlement they respond to is a global phenomenon that involves Australia. At the same time, displacement as method is a political sensitivity that has a long Australian tradition.[2]

Seasteaders

Explicitly referring to 'homesteading', the nineteenth-century settler colonial practice of allocating 'free' farms to those who would build and reside in unclaimed public lands, the 'seasteading movement' advocates the establishment of 'permanent autonomous ocean communities' (the idea of permanently living at sea on fixed man-made structures, of course, is not recent, see O'Hanlon 2008).[3] Like many of the 'pioneers' of old, in what would become the contemporary settler societies, the seasteaders aim to escape locales they have in a way already abandoned. The seasteaders fear the possibility of redistributive taxation to the point that they have decided to act politically as if such a policy had already been enacted. While responding to a reality that is not there is as good a definition of delusional paranoia as any, seasteading constitutes a settler colonial reflex. Theorist and organiser Petri Friedman notes: 'When Seasteading becomes a viable alternative, switching from one government to another would be a matter of sailing to the other without even leaving your house' (Friedman quoted in Wilkey 2012).[4] It is a somewhat extreme proposition, but it is not isolated.[5] Among its supporters are the ultraconservative Cato Institute and billionaire libertarian and PayPal founder Peter Thiel. Former Special Assistant to President Ronald Reagan Doug Bandow is an enthusiastic supporter of seasteading. In an article entitled 'Getting Around Big Government' he provides a brief review of the main works emanating from the Seasteading Institute and concludes that 'Seastead advocates are not crazed anarchists against the government' (Bandow 2012). In an article for the Cato Institute, Thiel pointed out that as there 'are no truly free places left in our world [...] the mode for escape [sic] must involve some sort of new and hitherto untried process that leads us to some undiscovered country' (Thiel 2009). He believes that the internet may emerge as a 'free place' but funded the Seasteading Institute as well (which seems justified, considering how the internet is used these days as a site of repressive surveillance).[6]

Why would anyone want to colonise the ocean surface? Wayne Gramlich lists a few reasons but prefers one above the others:

> There are a number of reasons – adventure, religious freedom, tax avoidance, trying out new forms of government, etc. Of the ones listed,

tax avoidance is my pick as the most powerful motivator for the development of sea surface colonization technology.

(Gramlich n.d.)

Actual seasteading is not likely, especially because there is no real need to actually sail away to artificial islands in order to avoid paying taxes, already existing islands in the Caribbean and elsewhere are already effectively providing these services, but seasteading, as many of its supporters note, is more important as an articulation of a political idea than practical reality. While the Seasteading Institute's blend of anarchist bohemianism, neoliberal corporate aggression and settler pioneering rhetoric is not unprecedented, it is noteworthy that the organisation promotes a *modular* form of architecture. Individual seasteads must be mobile. As serial modularity and systematic displacement become the principal method of political organisation, the prospect of change through mobility confirms a fundamentally settler colonial imagination. Other traits that are typical of the political traditions of settler projects include a focus on immediately self-sovereign communities, autonomy and, as mentioned and most significantly in this context, the possibility of moving without shifting place. Settlers, after all, are not migrants who join someone else's country; they move towards *their* country.

Seasteading is not, however, only about simply displacing and establishing alternative political systems. It is meant to promote political transformation where the seasteaders are heading *and* where they are coming from. Its promoters have given up on reform, both constitutional and legislative, and would not entertain the prospect of engaging in political struggle. Displacement is their only available answer. Just like in Ayn Rand's *Atlas Shrugged* – and the Seasteading Institute refers to Rand's influential book through its logo depicting Atlas emerging from the ocean supporting a seastead – once the government's hostile 'takeover' of society has happened, the best course of action for the 'best minds' is to displace elsewhere until they are required back (Rand 1999 [1949]; Seasteading Institute n.d.). Seasteading promoters Petri Friedman and Brad Taylor do not trust proposals that 'rely on the reform of existing institutions or the consent of existing governments'. 'In a competitive market for governance', they note, 'we should expect governments to make such concessions; in the current uncompetitive system, we should not' (Friedman and Taylor 2012, 218). They propose to displace, not to fight:

the current monopolistic equilibrium requires us to focus on the non-institutional determinants of competition: the geographic and technological environment in which governments are embedded. To robustly improve governance, we need to intervene at this bare-metal layer *rather than attempt to directly reform existing policies or institutions*. We propose an unorthodox form of intervention which we argue would achieve this goal – developing the technology to create permanent, autonomous settlements on the ocean. Settling the ocean – seasteading – would open a new frontier. The freedom of international waters allows for the introduction of *new competitors into the governance market without reforming the old system.*

(Friedman and Taylor 2012, 219; emphases added)

This refrain is typical of the political traditions that historically advocate settler colonialism. It can be described as the 'patent model' for social change: build an exemplary polity elsewhere and people will emulate. After all, the 'city upon a hill' as a model of political organisation must be visible from the world that is left behind (FitzGerald 1981). The possibility of producing political change where the seasteaders are coming from is not lost on Friedman and Taylor: to 'robustly improve government' everywhere, they note, 'we need to promote competition by lowering the cost to consumers of switching governance providers' (Friedman and Taylor 2012, 222). The idiom may be ultramodern and neoliberal, but their words hark back to what New Zealand historian James Belich has called the 'settler revolution' of the nineteenth century (Belich 2009), when, freed from the colonial supervision exercised by imperial authorities, immediately sovereign collectives moved towards sites of constitutive transformation that did not require revolutionary upheaval.

Indeed, Friedman and Taylor have a very particular history in mind: the North American prairies, for example, but they think it is an admirable example, only became a resource after technological innovation rendered them potentially productive locales. So it should be with 'acquatory', a neologism aimed at the territorialisation of water surfaces:

A useful way of thinking about the frontier is as the point at which the net economic value of some resource becomes positive. The new technology of the railroad, for example, gave land in the American west positive value to non-Indians, bringing it within the frontier. This *allowed new settlements outside the reach of any state* and thus lowered barriers to entry in the governance market.

(Friedman and Taylor 2012, 223; emphasis added)[7]

For them, improving society requires a 'new frontier': 'a blank canvas on which social or constitutional entrepreneurs can create their products and test them in reality by seeing if they can attract citizens' (Friedman and Taylor 2012, 223–224). Taking into account a determination to not put up a political fight, all of this, in turn, requires a 'dynamic geography'. As always, the geographies of settler colonialism begin as 'blank' canvases, become 'territories' (eventually 'Territories'), and finally actual places:

fluidity lowers the costs of switching government. If a family owns its own floating structure and becomes dissatisfied with the government it belongs to, it can simply sail away to another jurisdiction: with dynamic geography, people can vote with their houses.

(Friedman and Taylor 2012, 224)

For the seasteaders, a dynamic geography and even 'polycentric law' (Mendenhall 2012), an extreme form of privatised legal practice, are much, much better than impossible reform or the violent overthrow of existing political structures. The seasteaders have considered these options, but decided against such courses: 'Unfortunately, political instability tends to be accompanied by bloodshed, producing a trade off between peaceful stability with high levels of rent-seeking and violent instability with low levels of rent-seeking', they note. On the contrary, they conclude, '*Seasteading allows us to have political instability without bloodshed*' (Friedman and Taylor 2012, 225; emphasis added). The seasteaders advocate a form of permanent settler revolution.

Tiny houses

A settler colonial reflex can be detected in recent attempts to recover a sense of control in the face of crisis by moving to new places as a way to adopt new practices. The 'tiny houses movement' is one such contemporary instance of displacement as method by which the act of moving house is linked to a transformed way of living. True, this movement has significant antecedents and 'back to the land' and 'self-sufficiency' responses have been frequent at least since the 1960s, and exhibits a leftist anti-modernism in contrast to the rightist libertarianism of the seasteaders. It is in many ways, however, a new development and it is significant that while 'tiny houses' have been proposed as a possible solution for homelessness, the movement's primary appeal is for the embattled middle classes seeking financial freedom and a smaller ecological footprint in the aftermath of the global financial crisis (see Ashford 2015; Dirksen 2012; Federico-O'Murchu 2014; Tortorello 2014; Wilkinson 2011).[8] Tiny homes are attracting a lot of attention. Oprah Winfrey took an early interest (Oprah.com 2007) and Portland (where else?) even has a 'tiny house hotel' (Tiny House Hotel 2015).

The beginning of this movement's most recent incarnation can be traced to Sarah Susanka's 1997 *This Not So Big House*. Resonating with a number of typically settler colonial refrains, Susanka emphasised the possibility of freeing people's time by opting out of the market and focusing less on earning money. An escape from contradictions that is premised on the physical displacement to a 'new' locale (even if in this case it is a mono-locale) is as good a definition of a settler colonial impulse as any (for architectural examples, see Kahn 2012; Paredes Benítez and Sánchez Vidiella 2010; Richardson 2011).

Founded in 2002, the Small House Society is 'a cooperatively managed organization dedicated to the promotion of smaller housing alternatives which can be more affordable and ecological' (Small House Society n.d. [II]). The option of emplaced transformation is in this case also rejected; like seasteaders, the tiny house people also 'vote' with their houses. Emphasising mobility, tiny houses are indeed often on wheels, a strategy that enables them to bypass zoning regulations that favours what members of this movement regard as overbuilding. Importantly, however, even if they look like them, they are not recreational vehicles (RVs). A RV, such as a caravan, is meant to look and feel as much as possible *like* a house without becoming one; it approaches but does not cross the threshold separating them. A 'tiny house', on the contrary, is meant to be as much as possible *unlike* a house without becoming something else. With the tiny house, an escape becomes a determination to permanently displace as a political act.

The 'tiny r(E)volution' 'manifesto' also underscores the need to respond to an unprecedented crisis. The emphasis is on a profound discontinuity with the past:

> Somewhere along the line the American Dream became defined by owning more stuff than your neighbor and having the best quality money could buy. Many times that meant relying on credit that was *unsecured* and came with lofty interest rates. But is that the way to go? Is that the new truth? Do we need a bigger house, a better car, or a large salary to find happiness? And just what is this elusive happiness anyway? Does it come about when we sacrifice our dreams for the pursuit of stuff?
>
> (Odom and Odom n.d.; emphasis added)

It is a somewhat circular logic dealing in dreams, the 'American Dream' and the ultimate sacrificing of 'dreams', but it is also about economic risk: 'unsecured' credit produces a situation in which real estate 'has become a risky investment'. Those 'who do own homes', the manifesto concludes, 'are seemingly stuck in a vicious cycle of working just to afford the home they currently have; homes that are often larger than needed' (Odom and Odom n.d.).

This movement typically emphasises flexibility, especially the ability of responding to natural or unnatural, man-made disasters: hurricanes, and financial meltdowns, for example (see Holtby 2013; Shafer 2009). In a world that has become 'unsecured', the search for financial and environmental 'security' produces an impulse to displace. A list of benefits associated with small homes issued by the Small House Society confirms a search for sustainability:

> People who live in smaller living spaces generally own fewer possessions, consume less, and have lower utility bills. Smaller homes require less building materials for construction and smaller land use – therefore costing much less to purchase, maintain, and live in. Construction of smaller homes can utilize more efficient, natural, healthy, high-quality materials that might not be affordable in a larger dwelling. All of these benefits result in healthier, more cost effective living, and a better environment.
>
> (Small House Society n.d. [I])

It is indeed the perception of an incoming crisis that prompts the determination to reduce current investments in *unmovable* assets. Mobility is crucial to this reasoning; owning less increases fluidity. Despite likely differences in their environmental politics, seasteaders would approve.

But there is more. Beyond the possibility of living rent-free and reducing debt, the movement also attempts to unilaterally reduce the fiscal burden. Occupying smaller areas, small houses pay less local taxes; that is why local neighbourhoods are often very concerned about tiny houses: real estate values may fall and a larger share of income may be needed to make up the shortfall. In this sense, like the prospect of seasteading, the tiny houses movement challenges existing sovereign arrangements. It does so at two crucial junctures: at the level of regulatory regimes, and at the level of rent extraction (it is in this sense that the movement often represents itself as advocating a 'revolution'). Jay Shafer (2009), one of its most articulate promoters, links the ability to reprioritise personal choices while bypassing zoning and building regulations and escaping an economy that has come to rely on debt and rent extraction.

It is a determination to programmatically *avoid* regulation rather than undertaking to transform or inform it via traditional political means that alerts us again to the movement's foundational reliance on displacement as method. In a recent article, Mona Chollet (2015) remarks on the ways in which the tiny homes movement fundamentally accepts that housing is expensive and endeavours to comply rather than challenge a housing regime that is ultimately shaped by very political decisions and economic interests. She quotes Shafer: 'the tangible happiness of a well-lived life is worth a thousand vehement protests'. For him, change is not linked to active political contestation. In some ways, like the Freemen on the Land I describe in the following section, the movement practices a type of 'lawful rebellion'.

However, irrespective of whether it constitutes a genuine instance of resistance against an increasingly dispossessory regime, responding to the crisis focuses on displacing somewhere else (i.e. into a small dwelling). That the solution that is prospected constitutes a recognisable settler colonial reflex should be noted. Besides, while it is a recent 'movement', currently enjoying a remarkable success, and there is even going to be a reality TV show in 2014 (*Tiny House Nation*; see FYI n.d.; Green 2014), the movement draws on a storehouse of typically settler colonial sensitivities. 'Living large' in small cabins or getting 'away from it all', Henry David Thoreau-style, are classic tropes of settler colonial representations (Walker 1987).[9] In all these representations, the accent is on a deliberate decision to displace. In *Walden*, Thoreau had noted:

> I went to the woods because I wished to live deliberately, to front only the essential facts of life, and see if I could not learn what it had to teach, and not, when I came to die, discover that I had not lived.
>
> (Thoreau 1854, #16)

Living deliberately then required displacement. While there is a long environmentalist tradition linking small housing, wilderness, social isolation and small social units (the tiny houses movement focuses on individuals and couples – Jay Shafer, one of the movement's most prominent advocates, moved out of his tiny home when he built a family; see Chollet 2015), the tiny houses movement is proposing to live *deliberately* now by moving somewhere else. Of course, displacement these days does not need to be towards literal 'woods'. Tiny houses can fit in urban backyards and, at the other end of the spectrum, the availability of internet 'frontiers' shapes in crucial ways the design of the tiny houses.[10]

The Freemen on the Land

In 2012 CBC TV aired a report on the 'Freeman on the Land'. The report estimated this movement's membership in Canada at around 30,000 and mentioned some of their activities: tax avoidance, claiming a 'common law' right to the land, issuing their own car registration plates and treating their birth certificate as security (CBC News 2012). *The Economist* published an article focusing on the Freemen on the Land movement and its global growth in 2013. It linked the movement with the 'disorientation' that followed the global financial crisis, emphasised their ostensible insanity and, noting its presence in the United States, Canada, Britain, Ireland, Australia and New Zealand, remarked on its transnational nature (*The Economist* 2013). These are societies that share a common legal tradition, and the movement is a phenomenon that is typical of the 'Anglosphere', but these are also settler societies, or societies that have a significant tradition of *settler* emigration. In important ways, the emigrants from the British Isles went to the settler colonies as co-ethnic settlers endowed with inherent rights, rights that other migrants could not access.

Like seasteaders and the tiny houses people, the Freemen on the Land express an impulse to displace. The movement is indeed premised on the fundamental notion that it is possible and legal to *opt out* of society. As 'commonly known as Dom' noted in relation to court proceedings, '*if you don't consent to be that "person", you step outside the system*' (quoted in Wagner 2011; emphasis added). 'Stepping outside', of course, implies a form of displacement. They may not aim to move to international waters, or to houses on wheels, but the Freemen's reflex is fundamentally similar to that of the other movements considered here. The Freemen release their legal teams, deny the courts' jurisdiction, ask the judge if their authority relies on maritime law, read esoteric sentences in Latin and state that they are the 'official representative of the legal fiction known as' himself or herself (see Gardner 2011; Kent n.d.). They believe that all interactions between the state, the courts and individuals are contracts, and use strange names in an attempt to separate the person from the legal entity representing it (Salzyn 2013, 226–228).[11] This last detail produces indeed another type of displacement. The best summary of their legal theory that I could find is RationalWiki's:

Freemen believe that an individual has two personas. One of them is a physical, tangible human being, and the other is their legal person, personality, or strawman: a legal fiction created when a birth certificate is filed with what would normally be considered someone's name (*e.g.*, JOHN SMITH), capitalization being a particular obsession. They believe their birth certificate *is* their legal person, and will attempt to present it in court when said person is called for, rather than identifying themselves as that person. Freemen believe that all legal actions, restrictions and statutes can only be applied to their legal personality, and that, by separating themselves from their legal person, they can free themselves of having to abide by statute laws they don't like (or acts, as they would insist they are not laws).

(RationalWiki n.d.)

Beyond displacements, like seasteaders and like the tiny house people, the Freemen on the Land also challenge existing sovereign arrangements:

Freemen see a distinction between (what they call) common law and statute law, which they refer to as 'admiralty law', 'law of the sea', 'maritime law', or the 'universal commercial code' (a distortion of the US-only *Uniform Commercial Code*) – something that only applies to corporations, *e.g.* legal persons, not flesh-and-blood humans. They see admiralty law as being the law of commerce, the law of owner*ship*, citizen*ship*, and indeed anything else ending in '-ship'. They see evidence of this in various nautical-sounding terms used in court, such as 'dock', 'birth [berth] certificate', '-ship' suffixes and any other fancy word they think might have a vaguely naval sound. Freemen will take this further by referring to the court as a 'ship', its occupants as 'passengers' and claiming that anyone leaving are 'men overboard'.

(RationalWiki n.d.)

Like seasteaders and tiny house people, the Freemen on the Land movement also has significant antecedents. It uses legal concepts from the Sovereign Citizen and other right-wing tax protester movements, but expands on them, incorporating concerns that are typical of other right-wing movements like the Posse Comitatus (i.e. they believe that power should reside in the county and not with more distant units of government). Its understanding of the common law and its preoccupation with various theories on finance are similar to those of the Montana Freemen and other Sovereign Citizen movements of the 1970s and 1980s. The Freemen on the Land, however, are less political, and some see them as a 'resistance community'.[12]

Their ascendancy, however, cannot be appraised without reference to crisis. Dreaming of rights they may retain, however, the Freemen on the Land ultimately focus on their powerlessness. As one commentator noted, their rhetoric is particularly appealing to 'desperate, vulnerable people who are going through terrible times in their lives' (Wagner 2011). Most importantly, the movement is appealing to people who are losing control over their lives and property. The Freemen on the Land are especially concerned with debt recovery issues, and to those who would consider (metaphorically and psycho-legally, as we have seen) displacing.[13]

The Freemen on the Land focus on personal sovereignty, on their understanding of the common law, and on a supposedly betrayed 'ancient' (and unwritten) constitution. They see governments as corporations attempting to involve them in unwanted contracts. As Stephen Kent summarised (n.d.), a Freeman would argue that 'he or she has no obligation unless the litigant has explicitly formed a contract for that obligation'. This is another typical trait of settler colonial imaginings. 'Frontier' circumstances are defined by the temporary absence of state institution. Corporate structures typically compensate. But while corporate forms of governance informed the settler polities their ancestors lived in or joined, current dispensations validate the Freemen on the Land esoteric madness in other ways. They are able to see governments as corporations because corporations are these days often endowed with quasi-governmental capacities in ways that would have been inconceivable a few decades ago.

The Freemen on the Land do not focus explicitly on 'home' like the seasteaders and the tiny home people do, but theorise the ownership of their homes in ways that challenge existing sovereign arrangements. After all, they are Freemen on the Land, not Freemen in a two bedroom apartment. In their case, this relation takes a synecdochical form: it is the Freemen's putative residency on the land that defines their political self-representation and crucially implies a home (a residence). It is that imagined 'home' and the increasing distance between feeling 'at home' and the perception of deteriorating circumstances that enables them to think their opposition. But if the seasteaders need to think about moving to spaces beyond the authority of governments, the Freemen on the Land think of places beyond the authority of the courts and corporations.

To protect themselves from corporate encroachment, the Freemen on the Land theorise the oxymoronic concept of 'lawful rebellion'. This notion of escaping from political sovereignty rather than challenging it links this movement with other settler colonial traditions, with displacement as method and with the two other movements discussed in this chapter. 'Lawful rebellion' as a type of nonrevolutionary and nonemplaced rupture, leads the Freemen on the Land towards imagining other displacement. The nautical obsession of this movement should be linked to the need to emphasise distantiation. The Freemen on the Land have supposedly opted out and rediscovered their personal sovereignty, and the notion of a metaphorical 'ship' and of a body of water separating their persons from the law/governments/corporations they have seceded from enables them to understand separation in a spatial way.

In this sense, they also move 'home' without moving, and to do so need to imagine the displacement of the sovereign institutions they contest (they are

'Freemen' *on the land* and only recognise a law that operates at sea – a fluid conception of sovereignty directly connects them with the seasteaders, even if unlike the seasteaders they see the sea as a space of unfreedom). In a fundamental way, the 'ships' of the Freemen on the Land operate like seasteads: they are devices that produce a separation between locales where existing sovereign arrangements apply and locales where one can opt out of existing governance structures. An emphasis on separation through the imagination of an original body of water (that is, after all, what settlers traditionally face) enables the Freemen on the Land to respond to the dispossessions/accumulations they are subjected to. Thus, if separation, as Massimo De Angelis seminally remarked (2001, 2004), is the constitutive feature of primitive accumulation, the Freemen on the Land are thinking about reverse separation. They are especially interested in reciprocity: they issue 'fee schedules' against the government, demanding payment for their time, produce massive quantities of pseudolegal documents, and engage in what judge Rooke defined as 'paper terrorism' (2012).

Moreover, the 'Freeman on the Land' movement cannot be understood without reference to the settler colonial situation and a search or indigeneity.[14] Its name refers to the *villani* and other tenants in preconquest England (the Indigenous inhabitants that would be vanquished by Saxon invaders). Reference to a 'preconquest' time is key to the Freemen's reclamation of sovereign capacities. This is an ancient structure of feeling, fuelling a rhetorical tradition that was especially strong during the nineteenth century in the British Isles and in the British diaspora in other continents.[15] It is crucial that the natural law rights contained in this imagined constitution are portable and that in this context political change is imagined as a return and not as 'progress' or as a forward movement. Reclaiming an unsurrendered sovereignty, however, is also a typical call of Indigenous movements in the settler societies the Freemen on the Land operate in.

Indeed, indigeneity is key to these formulations (the Freemen even recently advised Aboriginal Australians on ways to obtain/withhold recognition – see Glazov 2014). Many of the groups analysed by Stephen Kent claim to be unconquered Indigenous peoples. Kent refers to the 'followers of the Moorish Law community' and the 'Moorish Nation', the 'United Mawshakh Nation of Nuurs', and the 'Washitaw Nation/Empire'. It is significant that the latter group claims to descend from the ancient moundbuilders of the Mississippi Valley, a group that can claim an indigeneity that is prior to that of really existing Indigenous groups (Kent n.d., 8, n. 4). Kent also cites Justice Rooke in relation to other 'indigenisations': 'black [M]uslims who self-identify as "Moors"' claim that they 'are not subject to state or court authority because they are governed by separate law, or are the original inhabitants of North and South America' (Kent n.d., 8). If the seasteaders and the tiny house people want to live deliberately *like* the settlers of old, the Freemen on the Land are settlers who want to be treated *unlike* the Indigenous peoples (of old and of today).

To practice their disassociation, the Freemen on the Land ultimately disassociate from society. Do they disassociate psychically as well? Jennifer Pytyck· and Gary Chaimowitz (2013) have taken seriously this question and focused on whether they are actually fit to stand trial. The Freemen on the Land *look* mad. The disassociation they practice between their flesh and blood and their legal *personae* is in important ways internal to their being. Indeed, the Freemen on the Land utter statements that appear as 'psychotic co-optations of established legal terminology', and an analysis of their 'lexicon' confirms that in their attempts to 'withdraw their consent to be governed', they ultimately 'repudiate their identity' (Pytyck and Chaimowitz 2013, 150–151). Pytyck and Chaimowitz also focus on other aspects of the Freemen's beliefs: that an original basis for legitimate government was corrupted, that as it was abandoning the Gold Standard in 1933 the government began backing the value of the American dollar by using its citizenry as collateral and issuing birth certificates as representations of that person's 'value' (those days, like these ones, were also marked by what was perceived as unprecedented economic crisis), and that 'redemption' enables a division of their 'flesh and blood' selves from their 'corporate entities or "strawmen"', but conclude that they are not actually mad (Pytyck and Chaimowitz 2013, 152). Their delusions are supported by 'subculturally-normative beliefs', they conclude (p. 153). This should not surprise. If the role of ideology is to naturalise power and to make it appear to be beyond the possibility of human determination, all challenges to the sovereign order are instances of (political) madness.

Conclusion

The movements discussed here rely on the internet to propagate themselves. But this is not all; these ideas are available elsewhere. They are available, I'd like to suggest, even if proving this proposition may be impossible, in a cultural storehouse of ideas and structures of feeling that are intuitively accessed and easily mobilised. They rely on tropes that are deeply embedded in the political imaginaries of the settler societies. These structures of reference constitute the settler colonial 'archive', an archive that fundamentally shapes the settler colonial present (Veracini 2008).

Accessibility and resonance cannot account, however, for these movements' diffusion. While these movements link domestic sovereignty and political sovereignty in ways that especially resonate in political settings shaped by settler colonialism, where 'home and sovereignty are intimately connected, they do so during times of unprecedented crisis. Displacement as method is often revitalised during times of crisis. In the context of these political traditions, rising contradictions are not addressed in place and the search for new liberatory geographies is offered as an alternative to political struggle.

The focus is on the possibility of finding a new 'home' politically and literally. With the seasteaders, the tiny house people share an explicit focus on housing; with the seasteaders, the Freemen on the Land share a determination to challenge established sovereign orders. With the tiny house people, the Freemen on the Land resist a dawning 'creditocracy' (Ross 2013). Yet again, the Freemen on the Land challenge those who would repossess their homes, just as the tiny houses movement seeks to evade housing regulation.

The 'social movements' outlined in this chapter are attempts to unbound housing from wider regimes of political sovereignty through displacement as method. They seek to avoid and (as a result) re-work the regulatory and financial controls surrounding owner-occupied housing in market settler societies. Through their advocacy for displacement such movements aim to rework the relations between residents, property and housing. However, as they adopt and reassert a profoundly settler colonial 'common sense', the chapter shows that these displacements ultimately fail to challenge property controls and existing structures.

Notes

1 The supporters of these movements are not 'preppers'. They do not seek isolation and self-sufficiency to individually survive the catastrophic dissolution of constituted order; on the contrary, they seek to exercise their collective constituent capacities. On the 'Doomsday Preppers', see, for example, National Geographic's incredibly successful reality TV series (2011–).

2 'Tree changing' and 'sea changing', for example, should be considered in this context (see Burnley and Murphy 2004; McManus and Connell 2011).

3 'Permanent autonomous ocean communities' is the Seasteading Institute's definition of seasteading (see Friedman and Gramlich 2012; Friedman and Taylor 2012). On seasteading, see also Steinberg *et al.* (2012) and Sydell (2012). Sydell refers to a group of 'rich techies in Northern California', a mix of 'geeks and hippies'. This specific social milieu has been a hub of ideological neoliberal renewal since the 1970s.

4 It is perhaps significant that Petri Friedman is nephew of neoliberal champion Milton Friedman.

5 On islands as privileged sites for political experimentation, see Gillis (2004).

6 Thiel founded PayPal 'as an attempt to create a web-based currency that would undermine government tax structures' (see Steinberg *et al.* 2012, 1537).

7 Fluidity as relative lack of attrition was indeed a crucial factor in the development of the settler revolution and Belich (2009) emphasises the link between the 'transport' and the 'settler' revolutions. But while the transport revolution has been interpreted as part of modernity's drive to annihilate space, it is significant that, for the settlers, moving to the periphery was often about reinstating separation.

8 Dirksen's successful documentary's title captures this movement's self-constitution as a sovereign collective.

9 Primarily a construction manual, *Tiny Houses* begins with a section dedicated to 'Tiny Historic homes'. They include the 'English Settlers Cottage' in Plymouth, Massachusetts, the 'Thomas Jefferson's Honeymoon Cottage', the 'Frontier Cabin', and 'Henry Thoreau's Cabin'. Kirsten Dirksen's 'We the Tiny House People' also concludes with a reference to Thoreau.

10 Sarah E. Thorne, for example, has examined 'the coincidence of the "small house" movement with the transformation of the house into a "media centre"', and noted how 'digital media technologies have opened up a new virtual world

to explore that radically defies and blurs our conventional understanding of interior and exterior spaces' (Thorne 2014).

11 Salyzyn focuses on 'self-represented litigants' in Canada, including the Freemen, and outlines Justice John D. Rooke's classification of 'Organized Pseudolegal Commercial Arguments' litigants (2012). Stephen Kent (n.d.) also relies on Justice Rooke's work to outline the Freemen's activity, origins and ideology.

12 Mona Lilja and Stellan Vinthagen (2014) include the Freemen on the Land in their work on communities that 'try to avoid integration with biopower through the creation of *autonomous conditions* for alternative self-making' (120). The Freemen on the Land also participated in the British Occupy movement, albeit controversially (see Kent n.d., 12).

13 Stephen Kent (n.d.) links the growth of the Freemen on the Land movement with periodic rural crises: the rural crisis in the United States during the first half of the 1980s, the interest rates spike in Canada between 1978 and 1981. Recent rural crises in the UK and drought-stricken Australia are also mentioned. Of course, the recent rise of the Freemen on the Land is linked to the consequences of the global financial crisis after 2008 (3.9 million foreclosures in the US alone, affecting about ten million people). This is not only about rural crises, however. The 'detaxers' among the Freemen on the Land are often failed professionals.

14 On settler colonialism as fundamentally structured by settler claims to indigeneity, see Veracini (2010).

15 On the role of this rhetorical tradition in the constitution of Australian political history, see McKenna (2012).

References

Ashford, J. 2015. Live tiny, save big. *BBC News*, 28 June. Retrieved on 27 July 2015 from www.bbc.com/capital/story/20150626-a-new-status-symbol-less-space.

Bandow, D. 2012. Getting around big government: the seastead revolution begins to take shape. *Forbes*, 30 July.

Barnes, A. 2015. Melbourne-made documentary Small is Beautiful hits the big screen. *The Age*, 12 March.

Belich, J. 2009. *Replenishing the Earth: The Settler Revolution and the Rise of the Angloworld*. Oxford University Press, Oxford.

Burnley, I.H. and Murphy, P. 2004. *Sea Change: Movement from Metropolitan to Arcadian Australia*. University of New South Wales Press, Sydney.

CBC News 2012. Freemen movement captures Canadian police attention, 29 February. Retrieved on 20 July 2014 from www.cbc.ca/news/canada/story/2012/02/29/freeman-movement-canada.html.

Chollet, M. 2015. Les temps des claustrophiles. *Le Monde Diplomatique*, May. Retrieved on 25 July 2015 from www.monde-diplomatique.fr/2015/05/CHOLLET/52942.

De Angelis, M. 2001. Marx and primitive accumulation: the continuous character of capital's 'enclosures'. *The Commoner*, 2.

De Angelis, M. 2004. Separating the doing and the deed: capital and the continuous character of enclosures. *Historical Materialism* 12(2): 57–87.

Dirksen, K. 2012. We the tiny house people. Retrieved on 19 July 2014 from www.youtube.com/watch?v=lDcVrVA4bSQ.

Duke, J. 2015. Floating cities may be a reality by 2020. *The Age*, 11 April.

The Economist 2013. Freeloaders on the land. 12 October 2013. Retrieved on 16 October 2013 from www.economist.com/news/americas/21587804-american-style-anti-government-eccentrics-take-root-canada-freeloaders-land.

Federico-O'Murchu, L. 2014. Tiny houses: a big idea to end homelessness. CNBC. com. 29 March. Retrieved on 30 July 2015 from www.nbcnews.com/business/real-estate/tiny-houses-big-idea-end-homelessness-n39316.

FitzGerald, F. 1981. *Cities Upon a Hill*. Simon & Schuster, New York.

Friedman, P. and Gramlich, W. 2012. Seasteading: a practical guide to homesteading the high seas. Retrieved on 02 May 2014 from http://seasteading.wpengine. netdna-cdn.com/wp-content/uploads/2012/03/full_book_beta.pdf.

Friedman, P. and Taylor, B. 2012. Seasteading: competitive governments on the ocean. *Kyklos* 65(2): 218–235.

Fry, T. 2011. Urban futures in the age of unsettlement. *Futures* 43: 432–439.

FYI. n.d. Tiny House Nation. Retrieved on 22 July 2014 from www.fyi.tv/shows/tiny-house-nation.

Gardner, C. 2011. The law is not the enemy of protest but an essential tool of impartiality: a rejection of the legal apparatus by some 'freemen' Occupy protesters will only make social progress and justice impossible. *Guardian*, 17 November 2011.

Gillis, J.R. 2004. *Islands of the Mind: How the Human Imagination Created the Atlantic World*. Palgrave Macmillan, New York.

Glazov, R. 2014. Freemen movement targets Indigenous Australia: 'Sovereign citizens' are advising Indigenous Australians to operate outside the laws of the land. *The Saturday Paper*, 6 September 2014.

Gramlich, W. n.d. SeaSteading: homesteading the high seas. Retrieved on 17 July 2014 from http://gramlich.net/projects/oceania/seastead1.html

Green, P. 2014. So small but already a TV star. *New York Times*, 2 July.

Hill, C. 1972. *The World Turned Upside-Down: Radical Ideas During the English Revolution*. Penguin, London.

Holtby, M. 2013. *The Tiny House Revolution: A Guide to Living Large in Small Spaces*. CreateSpace, North Charleston, NC.

Kahn, L. 2012. *Tiny Homes: Simple Shelter*. Shelter Publications, Bolinas, CA.

Kent, S.E. n.d. Freemen, sovereign citizens, and the threat to public order in British heritage countries. Retrieved on 11 June 2014 from http://griess.st1.at/gsk/fecris/copenhagen/Kent_EN.pdf.

Lilja, M. and Vinthagen, S. 2014. Sovereign power, disciplinary power and biopower: resisting what power with what resistance? *Journal of Political Power* 7(1): 107–126.

Livingston, J. 2010. *The World Turned Inside Out: American Thought and Culture at the End of the 20th Century*. Rowman & Littlefield, New York.

McKenna, M. 2012. Transplanted to savage shores: Indigenous Australians and British birthright in the mid nineteenth-century colonies. *Journal of Colonialism and Colonial History* 13: 1.

McManus, P. and Connell, J. 2011. *Rural Revival? Place Marketing, Tree Change and Regional Migration in Australia*. Ashgate: Farnham.

Mendenhall, A.P. 2012. My 'country' lies over the ocean: seasteading and polycentric law. *Studies in Emergent Order* 5: 137–156.

National Geographic. n.d. *Doomsday Preppers*. Retrieved on 22 June 2014 from http://channel.nationalgeographic.com/channel/doomsday-preppers.

Odom, A. and Odom, C. n.d. Manifesto. Retrieved on 15 July 2014 from http://tinyrevolution.us/about/manifesto.

O'Hanlon, C. 2008. Seeland. *Griffith Review*, 20. Retrieved on 14 June 2014 from https://griffithreview.com/articles/sealand.

Oprah.com. 2007. Inside a 96-square-foot-home. Retrieved on 29 July 2015 from www.oprah.com/oprahshow/Inside-a-96-Square-Foot-Home-Video.

Paredes Benítez, C. and Sánchez Vidiella, A. 2010. *Small Eco Houses: Living Green in Style.* Universe Pub, New York.

Piterberg, G. and Veracini, L. 2015. Wakefield, Marx, and the world turned inside out. *Journal of Global History* 15: 3.

Pytyck, J. and Chaimowitz, G.A. 2013. The sovereign citizen movement and fitness to stand trial. *International Journal of Forensic Mental Health* 12(2): 149–153.

Rand, A. 1999 [1949]. *Atlas Shrugged: 50th Anniversary Edition.* New York, Penguin.

RationalWiki, n.d. Freemen on the land. Retrieved on 20 July 2014 from http://rationalwiki.org/wiki/Freeman_on_the_land.

Richardson, P. 2011. *Nano House: Innovations for Small Dwellings.* London, Thames & Hudson, Inc.

Rifkin, M. 2013. Settler common sense. *Settler Colonial Studies* 3(3–4): 322–340.

Rooke, J.D. 2012. Reasons for decision of the Associate Chief Justice J.D. Rooke. In Crystal Lynne Meads (Appellant) and Dennis Larry Meads (Respondent). Court of Queen's Bench of Alberta, 18 September.

Ross, A. 2013. *Creditocracy and the Case for Debt Refusal.* OR Books, New York.

Salyzyn, A. 2013. Canada: foreclosures, freemen, foreign law schools and the continuing search for meaningful access to justice. *Legal Ethics* 16(1): 223–229.

Seasteading Institute. n.d. Homepage. Retrieved on 15 June 2014 from www.seasteading.org.

Shafer, J. 2009. *The Small House Book.* Tumbleweeds Tiny House, Sonoma, CA.

Small House Society. n.d.[I] A voice for the small house movement. Retrieved on 22 June 2014 from http://smallhousesociety.net/about.

Small House Society. n.d.[II] Homepage. Retrieved on 19 June 2014 from http://smallhousesociety.net.

Steinberg, P.E., Nyman, E. and Caraccioli, M.J. 2012. Atlas swam: freedom, capital, and floating sovereignties in the seasteading vision. *Antipode* 44(4): 1532–1550.

Susanka, S. 2001 [1997]. *This Not So Big House.* Taunton Press, Newtown, CT.

Sydell, L. 2012. Don't like the government? Make your own, on international waters, *NPR Radio*, 17 December. Retrieved on 17 July 2014 from www.npr.org/blogs/alltechconsidered/2012/12/17/166887292/dont-like-the-government-make-your-own-on-international-waters.

Thiel, P. 2009. The education of a libertarian. *Cato Unbound: A Journal of Debate*, 13 April. Retrieved 30 July 2015 from www.cato-unbound.org/2009/04/13/peter-thiel/education-libertarian.

Thoreau, H.D. 1854, Where I lived, and what I lived for. Retrieved on 19 June 2014 from http://thoreau.eserver.org/walden02.html#14.

Thorne, S.E. 2012. The cleaving of house and home: a Lacanian analysis of architectural aesthetics. MA Dissertation, University of Western Ontario. Retrieved on 15 July 2014 from http://ir.lib.uwo.ca/etd/1014.

Tiny House Hotel. n.d. Main page. Retrieved on 29 July 2015 from https://tinyhousehotel.com.

Tortorello, M. 'Small world, big idea', *New York Times*, 19 February. Retrieved 12 April 2014 from: www.nytimes.com/2014/02/20/garden/small-world-big-idea.html?_r=1.

Veracini, L. 2008. Colonialism and genocides: towards an analysis of the settler archive of the European imagination. In Moses, A.D. (ed.) *Empire, Colony,*

Genocide: Conquest, Occupation, and Subaltern Resistance in World History. Berghahn, New York.

Veracini, L. 2010. *Settler Colonialism: A Theoretical Overview.* Palgrave Macmillan, Houndmills.

Veracini, L. 2012. Suburbia, settler colonialism and the world turned inside out. *Housing, Theory and Society* 29(4): 339–357.

Veracini, L. 2015. *The Settler Colonial Present.* Palgrave Macmillan, Houndmills.

Wagner, A. 2011. Freemen of the dangerous nonsense. *Legal Week*, 16 November 2011. Retrieved on 17 July 2014 from http://ukhumanrightsblog.com/2011/11/15/freemen-of-the-dangerous-nonsense.

Walker, L. 1987. *Tiny Houses: How to Get Away from It All.* Overlook Press, Woodstock.

Wilkey, R. 2012. Seasteading Institute convenes in San Francisco: group fights for floating cities. *Huffington Post*, 4 June. Retrieved on 25 May 2014 from www.huffingtonpost.com/2012/06/04/seasteading-institute_n_1568951.html.

Wilkinson, A. 2011. Let's get small. *The New Yorker*, 25 July.

9 'The best house possible'

The everyday practices and micro-politics of achieving comfort in a low-cost home

Michelle Gabriel, Millie Rooney and Phillipa Watson

Introduction

In this chapter, we explore the diverse ways in which people achieve comfort within their homes and the possibilities for sustainable adaptation of low-cost homes. We are interested in the tensions that arise between competing desires for home comforts (warmth, security, affordability and belonging) and reduced energy use, with the latter itself being increasingly viewed as a source of personal satisfaction and comfort. Following social practice theory (Shove 2003; Shove *et al.* 2007, 2012), we understand the home as a socio-technical achievement, rather than a container in which social action takes place. This approach re-sets analytical distinctions between physical dwellings and social inhabitants, and instead traces the interweaving of human and nonhuman actors (heat pumps, furniture, pets and ornaments) in the process of achieving and managing the comforts of home (Shove 2003).

The findings presented in this chapter draw on research with residents who participated in a government-funded home energy-saving program, Get Bill Smart, in Hobart, Tasmania. We begin by discussing how comfort and energy practices are mediated by money in financially constrained households. We then present a sample of vignettes that illustrate the diversity of comfort priorities and how people's feelings of comfort are affected by the presence of other actors, human and otherwise. Here we report on competing comfort priorities and the dynamic nature of home living in order to make sense of the potential for and barriers to adapting dwellings and practices within the home. Following Head *et al.* (2013), we identify specific sites of 'traction' (i.e. opportunities for change) and 'friction' (i.e. resistance to change) that we observed when engaging with low-income households about home energy saving. The stories we report on contribute to an understanding of why it is that a particular energy-efficiency installation might work well in one home, but be disruptive or ineffective in another.

Finally, we reflect on the discord we observed between the Get Bill Smart project team's architectural and environmental aspirations for the 'best house possible' (a phrase used by one of our participants, Susan, to describe her home) in terms of thermal performance and energy efficiency, and the lived experience of the families who were making 'the best house possible' under constrained financial circumstances. Here we identify tensions between, but also possibilities for greater integration of, environmental and social outcomes associated with dwelling maintenance and adaptation.

Unpacking home comfort

There is now a substantial and diverse literature on home energy consumption and home comfort (for example: de Dear and Brager 1998; Guy and Shove 2000; Li and Lim 2013; Preval *et al.* 2010; Rosa *et al.* 1988; Shove 2003; Shove *et al.* 2010, 2014). Much of this work has focused on defining and measuring the phenomenon of thermal comfort through objective data on indoor house temperature and subjective data that document building occupant expectations around comfort. For example, de Dear and Brager (1998) advocate an adaptive thermal comfort model, which emphasises temperature monitoring and incorporates contextual factors that modify a building occupant's expectations and thermal preferences. More recently, this building-occupant approach has been criticised for failing to account for heterogeneity in household composition, household practices and lifestyle preferences (Ellsworth-Krebs *et al.* 2015). In contrast, Ellsworth-Krebs *et al.* (2015, 102) call for domestic energy research to engage with interdisciplinary scholarship on the 'home', which recognises the temporality of home and its connections 'to emotions and relationships, as well as social and cultural expectations'.

Elizabeth Shove's extensive work on comfort and convenience provides important groundwork to this relational approach to home. Shove's (2003) social practice theory focuses on how interactions between technologies and people give rise to particular social practices. She traces the co-evolution of comfort practices and technological developments, emphasising how constructed ideas of comfort are sustained over time. Shove's (2003, 43–46) work has informed new research on household sustainability (Gram-Hanssen 2010, 2011; Lane and Gorman-Murray 2011; Strengers 2011; Strengers and Maller 2011). While research in this area is centred on everyday practices within the home, importantly, home is recognised as an 'interface' that connects intimate spaces and practices to wider institutional and political processes (Hawkins 2011, 70). Strengers (2010, 7) notes that 'practices are often misunderstood as relating only to what people do, or to what they say about what they do, rather than the ways in which these "doings and sayings" are constituted and interconnected'. Head *et al.* (2013, 351–352) use the term 'connected household' to emphasise how 'home spaces and the people who live in them are inextricably linked in the social, technological and regulatory networks that make up suburbs, cities, regions and nations.'

Consistent with social practice theory, we are interested in how people interact with their physical dwelling and technologies around the home, and the contestation that occurs within homes among a range of inhabitants over achieving comfort and energy reductions. Following Susan Smith (2008), we use the term micro-politics to describe a microcosm of 'political' activity that is occurring within the home, while also recognising that such activity is shaped by technological developments, financial markets and governmental activity beyond the home. We are interested in how this micro-politics affects adoption of energy-saving measures and the success or otherwise of home energy-saving programs and policies. A focus on micro-politics is valuable in identifying the relational aspects of home energy consumption (Ellsworth-Krebs *et al.* 2015; Gram-Hanssen 2010).

Sustainable home adaptation

Much household sustainability research is focused on understanding the capacity for adaptation, including technical innovation, renovation and retrofitting, and attitudinal and behavioural changes. Social practice theory has contributed to debates over sustainable home adaptation by highlighting how energy-intensive scripts for living become embedded in technologies, as well as the entanglements and interconnections that contribute to the stabilisation of resource-intensive habits and regimes. As Shove *et al.* (2012, 19) argue, 'patterns of stability and change are not controlled by any one actor alone'; instead there are a 'range of elements in circulation'. In their explication of a social practice approach, Shove *et al.* (2012, 14–15) argue that practices are re-made through the recombination of various elements including 'materials', 'competences' and 'meanings', and that the role of the social analyst involves tracing how particular combinations are sustained and broken over time.

Others interested in sustainable home adaptation have focused on the in/tractability of household norms that inform practices and negotiations between household members. Strengers' (2011) research on the impact of feedback technologies on energy and water consumption calls attention to the importance of understanding negotiable and non-negotiable practices within homes. While households might be willing to opt for small-scale actions that lead to minor reductions in energy and water usage, she observes that a clearer understanding of the norms and expectations that inform certain practices is required before fundamental shifts in consumption patterns are possible.

Similarly, Head *et al.* (2013) are interested in the capacity for behavioural change and effective policy interventions around sustainable water consumption. Here, they distinguish between 'zones of friction' that they observe within homes (i.e. sites that are more likely to be resistant to sustainable interventions) and 'zones of traction' (i.e. established practices on which sustainable interventions can build) (Head *et al.* 2013, 353). They argue that these concepts are valuable in identifying constructive spaces of policy intervention for environmental sustainability. Following Head *et al.* (2013, 359), we seek to pay close attention to the 'conflict, jostling and variability within households'.

Research methods

The insights detailed in this chapter are based on data collected by the authors for a government-funded home energy-saving project, Get Bill Smart.[1] Get Bill Smart supports low-income households in Greater Hobart, Tasmania, to reduce energy use and to improve thermal comfort in the home. There are 498 low-income households participating in this project. Energy bill and survey data were collected from all participating households. In exchange for their project participation, households received grocery vouchers and/or a free home education and energy-efficiency upgrade. Of these households, 60 participated in detailed monitoring of energy practices and technologies around the home. This involved a home visit from the project team in which qualitative interviews were conducted, home observations recorded and energy, temperature and humidity loggers installed.

Each home visit involved a structured interview around issues of thermal comfort and energy efficiency in the home. Participants were invited to explain not only how comfortable their home was, but how and why they found it comfortable or uncomfortable, how they managed temperature and energy efficiency and the changes they have made or would like to make to this end. Questions were also asked about the availability of thermal comfort and energy efficiency knowledge and skills within the participant's broader communities.

Project funding required that we report on the physical condition and makeup of the house. For example: orientation, building materials, hot water heater type, window coverings, energy-efficient light bulbs, floor coverings, air vents, window frames and more. To gather these data the research team had to physically move around both the inside and outside of the house. During the process of these observations, householders were able to speak to various parts of the home, explaining the quirks of the physical dwelling and the social interactions that took place with and within the structure. For example, as we walked out the back door of one participant's home she said 'See how the wind hits here, I built this wall to stop it rushing through the house'.

At the time of the interview external consultants were also on site, fitting the house with electricity, temperature and humidity loggers which remained in the house for a 12–15-month period. The placement of these data loggers required some consultation with the householder and as a result we were able to learn more about how the home functioned – for example, which rooms were coldest, whether the kids were likely to pull out the power cords, which beds had electric blankets and which heaters were never used. This process built on insights gathered through the interviews and home observations, facilitating discussions about how the home functioned technically and socially.

In collecting the data for this research we have sought to integrate multiple types of data, both qualitative and quantitative from each of the homes. As Foulds *et al.* (2013, 627) have previously observed, the use of both types of data 'provides the depth required to reflect suitably on data collection, theoretical application and analysis-related issues'. While presenting relevant survey results and interview quotes about householders and their homes, we also present a selection of vignettes based on researcher observations in the field to tell the story of particular home situations. This case-based approach has recently been used by Gram-Hassen (2010) to understand variation in residential heat comfort practices and energy use across households. By pulling together quantitative and technical aspects of the home in the context of personalised stories and dwelling experiences we are able to, as Ellsworth-Krebs *et al.* (2015, 100) suggest, 'adopt the home (and all the baggage the term comes with) as the focus for investigation, highlighting an appreciation for the socio-technical nature of domestic energy demand'. The cases we describe in this chapter illustrate the micro-politics, tensions and negotiations that occur between inhabitants as they seek to make a warm, cosy home.

Findings

Turning to our findings, we begin with a discussion of energy bills. This is important as money is a major mediator of decision-making in financially constrained households. However, our time spent with participants made us aware that comfort (in its various forms) is constrained and enabled through other non-financial considerations (many of which were unexpected) within the home, including: draughts and mould; pet ownership and care; tenure arrangements; and memories.

Energy bills: an unwelcome guest

Long-term financial hardship constrains both the capacity for low-income households to pay their energy bills and their capacity to make substantial changes to the home that might potentially reduce such bills. However, concerns about cost can also drive practices around the home that are directed towards energy saving. Previous social research on Australian household sustainability highlights the complex and contradictory nature of household attitudes towards practices relating to home energy saving. In their study of Wollongong households, Gibson *et al.* (2011) noted that compared with high-income households, lower-income households generally expressed a lower commitment to climate change as an issue. However, for low-income households, sustainable household practices were more likely to be associated with resourcefulness and constrained consumption, rather than the purchasing of green technologies (Gibson *et al.* 2011, 30–31).

The Get Bill Smart project aimed to support low-income households who were potentially experiencing financial hardship. Therefore, a major focus of our work was to understand: participant concerns about energy bills; strategies used to pay their energy bills; and participant interest in reducing energy bills through energy-efficiency measures. Our survey data confirmed that participants were struggling to manage energy bills on a low income, with most participants (84 per cent, $n = 423$) indicating that they joined the project in order to save money. Over half of the participants reported that they had experienced financial hardship in the last 12 months (57 per cent, n = 286) and that they found it hard to pay their energy bills over the last year (55 per cent, n = 274).

Further, we found that financial constraints had a direct relationship to participant's ability to be thermally comfortable in their home. Energy use and temperature data indicated that there were a group of people who were foregoing warmth and comfort in order to save energy and, in turn, money. For example, Rachel had kept an accounts book of all her expenses, including her energy use, over the past 50 years. She knew that she only spent $1.50 each day on energy. In order to save money, Rachel did not heat her house. When we monitored her home during cold periods, we found that inside temperatures were often only one or two degrees warmer than outside. While low levels of energy consumption are desirable from an energy-savings perspective, the indoor temperature data indicated high levels of comfort deprivation.

Our discussions with participants also revealed significant levels of stress associated with energy bills. For example, one participant, Emily, used a function on her meter box allowing her to estimate the likely cost of the next bill. She explained: 'I check it normally once or so a week, just to see how it's going. I shouldn't because then it stresses me out, and I think oh no, we've used too much' (Emily, 4 July 2014). When it looked to be 'high', Emily would redouble her efforts to reduce power use within the home, encouraging her partner to turn off the television and refusing to turn it on during the day for her son. In Emily's situation, technology was enabling her to closely monitor her energy use, but because of financial hardship she had limited capacity to make change or to cover the costs of even low levels of energy use. Monitoring did not necessarily empower Emily, but rather fuelled her anxiety about pending bills, sometimes leading to tension between her and her partner.

In many situations, those responsible for the energy bills sought to change other family members' behaviours with varying degrees of success. For example, Tony lived with his wife and teenage children and called himself the 'light fairy'. He was constantly turning lights off in an effort to save power. Similarly he was frustrated by his children leaving their computers on when not in use and when he got sick of nagging he took action: 'I just yank the power cord, I don't shut them down, I just yank the power cord and if anything dies, tough luck.' Tony observed that energy reduction 'logic is simple, but it's not easy because you've just got the usual issues' (Tony, 16 October 2013).

Negotiating relationships: draughts, moisture and mould

Daniel Miller's research (2001, 2008) on material culture illustrates how the physical structure of a house actively shapes relationships and practices within the home. In the Get Bill Smart study the thermal performance of houses was generally poor. Many houses had little or no insulation, with less than half of the houses having comprehensive ceiling insulation (47 per cent, n = 236). Most houses failed to make effective use of solar orientation.

One house we visited had only one small window to the north and many large windows facing south. The owners of the home had changed the front door (on the northern side) from wood to glass to allow a little more light in, but they could not afford further renovation. Seventy-two per cent of houses (n = 362) surveyed had significant draught issues. The reasons for the draughts varied: a broken back door that a landlord refused to fix; cracked or missing windows; poor construction of the house; shrunken and ill-fitting timbers; and lack of insulation and sisalation. Windows were also often a problem; rattly aluminium or steel window frames that easily lose heat were common, as were poorly maintained wooden frames. In one house there was no insulation and the occupants could see their indoor lights through the roof when they stood outside at night. For Nancy, being in her house was like 'standing in a field' (Nancy, 1 October 2013).

In addition, most households (81 per cent, n = 401) observed moisture in their house, with around two-thirds of these households describing the level of moisture as either high (23 per cent, n = 90) or medium (43 per cent, n = 172); and a one-third of these households describing the level of moisture as low (35 per cent, n = 138). Around half of participants (49 per cent, n = 240) had noticed mould in their house over the previous year.

For some participants, the physical response of the house to climatic conditions shaped negotiations and trade-offs that took place. For example, Emily had noticed mould problems developing in her house and so made sure to always keep at least some windows open. This was despite her house already being extremely cold and draughty. Visiting her home provided us with a chance to observe the house firsthand.

Arriving at the house we're surprised by how new it looks. Most houses we visit in the study were built prior to the 1970s. This one shines as the sun bounces off the corrugated colourbond exterior. The house is rented by a family of four, with two children under the age of three. Constructed in 2012 this free-standing three-bedroom home was delivered, in two pieces, to the current site.

Like all houses built after 2003, on paper this one ticks the necessary energy efficiency boxes. Yet despite this it suffers from hot flushes in the summer and chills in the winter. On really hot days, the house finds itself empty as the occupants resort to driving around the suburbs in the comfort of their air-conditioned car, leaving behind only echoes of their heat induced grumpiness. In winter, the house struggles to hold heat. An inspection of the cavity above the laundry shows that although some insulation is present, it is poorly installed with large gaps that render it ineffective. The one heater wired into the wall works hard to heat the open plan living area and the large hallway that cannot be closed off.

Sometimes, when it's really wet it rains inside. Water cascades through the internal double doorway that separates the main living area from the hall. But even when it's not raining, in winter the house receives careful treatment as each morning the occupants wipe the condensation from not only the main windows in the house, but the walls in one of the children's bedrooms. Despite this care and attention to internal moisture, the house is still a site for mould. The mould grows not only on the original physical structure of the home, but on the belongings that have been placed inside it. Every couple of months the child's pillow is replaced due to mould inhabitation.

On our second visit to the house we notice that changes have been made. The wired-in heater is no longer used and instead it has been replaced by a plug-in heater. While the costs of running this heater will be higher, the occupants are able to position it so that the heat generated can be used more effectively in the living space and in the bedrooms. The occupants are trying various strategies, but they are yet to find a solution that can transform this house into an affordable, warm and healthy living space.

Negotiating relationships: dogs, cats and fish

In seeking to unpack the processes by which comfort is achieved in low-cost homes, we soon became aware of the entanglement of pets within this process. While a minor theme in household sustainability research, Emma Power's research on the role of domesticated pets and native animals draws attention to the presence of animals in the making of home. Power (2008, 537) calls for the need to consider pet-keeping cultures and to 'engage with more-than-human families' within social analyses of home sites. In relation to the domestication of dogs, Power (2012) observes that while owners seek to discipline their dogs to fit in with established family routines, the dogs simultaneously have a disruptive effect on domestic arrangements and practices.

Similarly, our encounters with family pets highlighted their multi-faceted effect on the home, with pets having both a disruptive effect on comfort habits and practices, as well as being a source of warmth and comfort for family members. We spoke to a resident and passionate pet owner who had carefully insulated her hot water tank only to find that her dog had ripped it off and spread it around the backyard. Her cats had also chewed through multiple electrical cords. We found that people's relationship with their pets shaped their capacity to make energy-efficiency changes around their home. While for some pet ownership and practices were negotiable, for others comfort achieved through the presence of loved pets trumped concerns about energy saving. For example, there were households in the study who had tropical fish-tanks, which require significant energy to run. On hearing how much it takes to run these tanks (sometimes up to $500 per year), one participant decided that they could not afford to keep the fish; another passionately declared it was worth the expense.

Patricia's story highlights the significance of pets in people's lives and how their presence impacts on household decision-making around energy use.

Patricia is a single 63-year-old woman who lives with a dog, cats, budgerigars, and some chickens clucking away in the backyard. Patricia lives in a single-storey, four-bedroom, detached house. She rents from a real estate agent who seems to care little for the upkeep of the place.

Patricia's house is cold, uninsulated and draughty. On windy days Patricia worries about the banging and shaking of the aluminium windows. The rattling is so great she's afraid the large panes of glass are going to break and shatter. Patricia explains that the primary source of heating, the wood fire, is unusable because it smokes out the living room. She says this has been a problem since she moved in, with the landlord refusing to clean the chimney. Smoke in the home is uncomfortable and exacerbates symptoms already experienced due to asthma and emphysema. Smoke aside, Patricia struggles to find money to buy wood. The wood heater is so inefficient that even when she has been able to afford the fuel, the heater has done nothing more than just take the chill off the air. As a result, Patricia rarely heats the house.

Patricia loves her pets and she prioritises their needs above her own. Patricia likes to keep the back door slightly open to allow her cats to come and go as they please. While her cats warm her lap in the evenings, leaving the back door ajar contributes to the sieve-like functioning of the house. When it is really cold Patricia just goes to bed and watches TV, sometimes she 'even has two dressing gowns on'. When asked whether this made her frustrated or depressed, Patricia replied that yes, in some ways it did, but that 'my animals' food comes first'.

Patricia considers herself to be on the poverty line, a place she ended up unexpectedly after business trouble. She has been struggling financially recently after one of her beloved cats was diagnosed with prostate cancer. Patricia is still paying $40 each fortnight from her pension to cover vet bills.

Negotiating relationships: other presences

In his work on material cultures within the home, Miller (2001) reflects on how a home's history and its decorative order can affect its occupants. For some, the materiality of the home is experienced as a constraint that 'haunts' the new occupant, whereas for others, the home is seen as an expression of the love of a family. Listening to the concerns of occupants in a council estate, Miller (2001) draws attention to 'invisible' presences within a home. Here he identifies the council as an 'uncaring and distant presence that possesses what the occupant cannot possess' (Miller 2001, 8). It is this other presence (the council landlord) which restricts home adaptation and redecoration within the council estate and, in turn, 'prevents people from feeling that their homes are an expression of their own agency' (Miller 2001, 8).

Our participants noted the invisible presence of the landlord, which influenced their interest and ability to make changes around the home. Tenure was viewed as a mediating factor in relation to capacity for change. For people like Jacob, the short-term nature of his tenure meant he had limited interest in making changes:

> There's probably a few minor things that could be done, but at the moment it's really up to the landlord to bring it up to standard really. He said he's probably going to end up selling it so I really can't see him investing in the time and money to bother fixing it.
>
> (Jacob, 20 July 2014)

Other people explained that their landlord was unwilling to make the necessary changes and unlikely to give permission for them to make the changes themselves. In some instances people decided to forego the chance of free upgrades available through the program as they did not want to destabilise the relationship they had with the landlord, despite the upgrades being financially beneficial for both the occupant and owner. The invisible presence of the landlord was not always negative. Caroline felt inspired to make changes to her rental home because she felt valued by her landlord. With permission, Caroline had renovated elements of the house and was pleased with the reaction of her landlord: 'She was so happy that I did it that she paid my water bill for me' (Caroline, 23 June 2014).

Other invisible presences were also significant for participants. In some houses, it was the memories of early family life that shaped their present actions within the home. The legacy of the past was not limited to the photos displayed along participant's mantelpieces, but rather for some the physical layout of the home was the material realisation of a former life. This material inheritance influenced a participant's disposition towards home 'improvement', with emotional connection to a pre-existing order inhibiting potential for change. For Susan, the memories of her late husband lived on in the house, which her husband built and worked on throughout their life together. While she wanted a warmer home and lower energy bills, she was reluctant to make any substantial changes to his handiwork.

When we spoke to Susan on the phone she explained that she didn't need anything done to her house as her late husband provided her with the 'best house possible'. Susan is in her mid-seventies and has lived for many years in the house her husband built. It was with great love and pride that she explained how he had always looked after her, created a wonderful home for them to live in and even provided her with 'all the mod-cons'; Susan was particularly pleased with the dishwasher. Despite her husband having passed away many years ago, Susan admitted that she often talks to him in the house. For Susan, any kind of significant adaptation to the house was unthinkable; her house was the remaining loving embrace of her husband.

Discussion

Achieving 'the best house possible' was a goal shared by many participants in our study. However, what made the physical houses and the experiences of the occupants different was that each of these homes was a unique composition of occupants, materials, arrangements and practices. In their study of household sustainability, Gibson *et al.* (2011, 26) observe that: 'The environment is just one line of responsibility being juggled in acts of consumption, which necessarily serve different anticipated needs'. Gibson *et al.* (2011, 28) also note that 'as the structure of households and material spaces of co-habitation change so too do the ethics and pragmatics of social worlds begin to shape decisions about resource use, sharing and consumption'. We also found that people's homes are dynamic, as illustrated by the arrival and departure of guests, the reorganisation of furniture and changes in family schedules. While some participants were dissatisfied with the physical layout of their homes, such as Tony, who wanted to reorganise the house to reduce noise and draughts, others, such as Susan, viewed their home as perfect just the way it is.

The program managers of Get Bill Smart sought to help participants achieve 'the best house possible', with 'best' denoting the 'most energy efficient' house possible. While the program aimed to reduce energy consumption, it was recognised by program managers that energy consumption is typically lower among low-income households than affluent households. The major problem faced by low-income households is that the energy they consume is not readily translated into improved warmth and comfort due to leaky, poor-quality housing. A $300 heating bill for a poor-quality home has very different comfort and implications to a $300 heating bill for a well-sealed, well-insulated home. In addition, the costs associated with such energy waste are not insignificant for low-income households, but rather they are a source of anxiety. Among low-income households the gains made through an energy-efficiency program are more likely to improve comfort and energy efficiency, rather than reduce overall energy use. Within existing home energy research this is referred to as 'take back' (Milne and Boardman 2000).

Notably, Get Bill Smart offered residents an opportunity to reduce their energy bills and potentially improve the thermal performance of the house. In Head *et al.*'s (2013) terms, there was considerable 'traction' between the levels of need among low-income households and the program objectives. People living in poorly performing homes and who were experiencing substantial discomfort in their home (such as Emily and Patricia) were more likely to be responsive to the types of energy-efficiency changes being promoted through Get Bill Smart. Participants identified money as a significant source of stress in their lives and also as one of the major barriers to change within their homes. The program went some way in addressing this through the provision of a free energy-efficiency upgrade, which included insulation, draught-proofing and efficient lighting.

In addition to minor home upgrades, householders were provided with information about low-cost strategies for reducing energy use during a visit from a home energy helper. Such strategies resonated with a range of participants, particularly those who expressed an ethos of resourcefulness and an aversion to wastefulness, such as Tony. There were also people such as Rachel with excellent record-keeping and budgeting skills, who had experience in closely monitoring their energy use. In these situations, home energy-efficiency programs can tap into and extend this existing interest and skill base. However, as Emily's experience illustrates, advice and technologies that support low-income households to become more disciplined about monitoring their energy use can also potentially fuel anxiety about energy bills. Programs that offer increased monitoring without resources for improving the thermal performance of the house are therefore problematic for low-income households.

While we observed many positive sites of traction between the Get Bill Smart project and the homes studied, there were also substantial barriers to delivering reductions in energy use and improvements in comfort. The scope of the Get Bill Smart project meant that many householders received the benefit of insulation and draught-proofing. However, for others, in more dire housing situations, these measures (which are recognised as delivering significant energy reduction) did little to address an extensive web of home energy and maintenance problems. Emily's house was draughty and, despite claims it was insulated, struggled to retain heat. One solution was to install draught excluders around doors and windows. In Emily's case, however, reduced air flow was likely to exacerbate the mould problem in one of the bedrooms. The mould posed significant health concerns for Emily as an asthmatic. Emily was also concerned about the mould that was reappearing on her son's pillow. She was trying to manage this problem by regularly replacing the pillow. At the house level, one of the few possible solutions to this mould problem would be to simultaneously heat the mouldy areas of the house and encourage the draughts. While this would address the health risks in the bedroom it would also dramatically increase the energy bills, which Emily already worried about and struggled to pay, without improving the thermal comfort. There is little that Emily can do to change the thermal performance of her dwelling short of a complete re-build. The other alternative is to move on to a new tenancy.

Insecure tenure is also seen as a major barrier to instigating changes around the home, with owners more willing and able to invest than tenants (Gabriel and Watson 2012). While many private rental tenants involved in the study did raise insecurity of tenure as a major barrier to making change, there were some exceptions. The good relationship that Caroline had with her landlord meant she was able to financially and emotionally invest time into the home and make significant thermal comfort and energy-efficiency improvements. In contrast, Susan owned her own home, yet felt unable to contemplate making changes for thermal comfort or energy efficiency. For Susan, the relationship that she has with the house as embodying the memory of her late husband acts as a site of friction when it comes to making change. These examples remind us to look beyond commonly identified and expected sites of friction and traction to the more nuanced practices in which individual householders are enmeshed.

While both financial security and tenure have previously been identified as barriers within home energy-saving research, our study reveals a range of less obvious sites of friction which are constraining possibilities for adapting the physical dwelling and reconfiguring household practices. Indeed, common in the vignettes above is the importance of maintaining and nurturing valued relationships with people, animals and even the home itself. These relationships act as barriers to change. Patricia, and others, value the maintenance of quality relationships with their pets over changes for thermal comfort and/or energy efficiency. While living in a cold and draughty house is bad for Patricia's health, it is more important to her that the doors are left open (significantly contributing to the draughtiness of the house) to allow her cats to come and go as they please. Similarly, Susan prefers the comfort of living in her home just as her husband left it rather than making changes to improve the house's thermal comfort. These non-thermally related forms of comfort act as sites of friction in attempts to change energy efficiency and thermal comfort behaviours in the home.

Conclusion

While our experience of visiting and monitoring low-cost homes highlighted the pervasiveness in building dysfunction across the housing stock, closer inspection revealed diversity in people's responses and practices relating to achieving home comfort. Individual household responses were informed by diverse household priorities, household arrangements and relationships. As Tony pointed out, energy reduction 'logic is simple, but it's not easy because you've just got the usual issues'. Here Tony is alluding to the dynamics of home life, which recent scholarship on the home has sought to unpack

(Ellsworth-Krebs *et al.* 2015; Lane and Gorman-Murray 2011). Homes are not physical containers in which social action takes place, but rather the home is unbounded, restless and indivisible from our emotional world and social and cultural expectations. Home comfort is achieved through a range of entanglements that can stabilise over time, but which are always open to being reordered within wider networks of environmental, political and cultural relationships.

The diversity of the homes we visited challenged the scope of the Get Bill Smart project and the predetermined energy-efficiency upgrade measures available through the program. The Get Bill Smart program sought to support people to achieve 'the best house possible' from an energy-efficiency perspective. Here a good, efficient house was understood to be a dwelling that has: good thermal resistance; appropriate and efficient heating system for the main living spaces; comprehensive insulation; energy-efficient home appliances and lighting; and residents who abide by certain energy-saving practices. While people wanted warm homes and there was considerable traction between the program goals and level of need, participants also sought a range of competing comforts such as maintaining good relationships with people, pets and presences both inside and outside the home, achieving financial control, and maintaining identities. Complicating the potential for Get Bill Smart to achieve its goals were the expected and unexpected sites of friction as detailed in our findings: the scale of home maintenance problems; home attachment and tenure; and family relationships.

Our experience highlights the need to look beyond commonly identified and expected sites of friction and traction to a wider suite of household practices. In particular, an understanding of the interaction between practices within a home is important in identifying appropriate interventions for individual houses. As Shove *et al.* (2012, 19) note, 'patterns of stability and change are not controlled by any one actor alone'. While the use of draught-proofing stems heat loss, reduced air flow can have consequences for mould growth. Also of significance is an understanding of the temporal and dynamic nature of home life, including changes in family schedules, changes in household composition, house moves and technological innovation. Such understanding highlights the limits of one-off installations, and the value of attending to home maintenance over an extended period.

Critically, our experience highlights the need for significant engagement with householders and their homes prior to identification of potential solutions related to home comfort. Such engagement is important in developing tailored, flexible responses that can accommodate existing relationships and attachments within the home. Understanding the relationships and micro-politics within and beyond the home is valuable in identifying and facilitating potential home energy interventions that are valued by the target household. This approach is a step towards effective and durable dwelling interventions that deliver social and environmental benefits; giving rise to sustainable houses and sustaining homes.

Note

1 Get Bill Smart is funded through the Australian Government's Low Income Energy Efficiency Program (2013–2016). The project is a collaboration between Sustainable Living Tasmania, Mission Australia and the University of Tasmania. The authors are responsible for the evaluation component of the project. The views expressed herein are not necessarily the views of the Commonwealth of Australia, and the Commonwealth does not accept responsibility for any information or advice contained herein.

References

de Dear, R. and Brager, G.S. 1998. Towards an adaptive model of thermal comfort and preference. *ASHRAE Transactions* 104(1): 145–167.

Ellsworth-Krebs, K., Reid, L. and Hunter, C.J. 2015. Home-ing in on domestic energy research: 'House,' 'home,' and the importance of ontology. *Energy Research and Social Science* 6: 100–108.

Foulds, C., Powell, J. and Seyfang, G. 2013. Investigating the performance of everyday domestic practices using building monitoring. *Building Research & Information* 41(6): 622–636.

Gabriel, M. and Watson, P. 2012. Supporting sustainable home improvement in the private rental sector: the view of investors. *Urban Policy and Research* 30(3): 309–325.

Gibson, C., Waitt, L., Head, L. and Gill, N. 2011. Is it easy being green? On the dilemmas of material cultures of household sustainability. In Lane, R. and Gorman-Murray, A. (eds) *Material Geographies of Household Sustainability*. Ashgate, Surrey, 19–34.

Gram-Hanssen, K. 2010. Residential heat comfort practices: understanding users. *Building Research & Information* 38(2): 175–186.

Gram-Hanssen, K. 2011. Understanding change and continuity in residential energy consumption. *Journal of Consumer Culture* 11(1): 61–78.

Guy, S. and Shove, E. 2000. *A Sociology of Energy, Buildings and the Environment*. Routledge, London.

Hawkins, G. 2011. Discussion: interrogating the household field of sustainability. In R. Lane and A. Gorman-Murray (eds) *Material Geographies of Household Sustainability*. Ashgate, Surrey.

Head, L.M., Farbotko, C., Gibson, C., Gill, N. and Waitt, G. 2013. Zones of friction, zones of traction: the connected household in climate change and sustainability policy. *Australasian Journal of Environmental Management* 20(4): 351–362.

Lane, R. and Gorman-Murray, A. 2011. Introduction. In R. Lane and A. Gorman-Murray (eds) *Material Geographies of Household Sustainability*. Ashgate, Surrey.

Li, D. and Lim, D. 2013. Occupant behavior and building performance. In R. Yao (ed.) *Design and Management of Sustainable Built Environments*. Springer, Reading.

Miller, D. (ed.). 2001. *Home Possessions: Material Culture Behind Closed Doors*. Berg, Oxford.

Miller, D. 2008. *The Comfort of Things*. Polity, Cambridge.

Milne, G. and Boardman, B. 2000. Making cold homes warmer: the effect of energy efficiency improvements in low-income houses. A report to the Energy Action Grants Agency Charitable Trust. *Energy Policy* 28(6–7): 411–424.

Power, E. 2008. Furry families: making a human–dog family through home. *Social and Cultural Geography* 9(5): 535–555.

Power, E. 2012. Domestication and the dog: embodying home. *Area* 44(3): 371–378.

Preval, N., Chapman, R., Pierse, N., Howden-Chapman, P. and the Housing Heating and Health Group. 2010. Evaluating energy, health and carbon co-benefits from improved domestic space heating: a randomised community trial. *Energy Policy* (38): 3965–3972.

Rosa, E., Machlis, G. and Keating, K. 1988. Energy and society. *Annual Review of Sociology* 14: 149–172.

Shove, E. 2003. *Comfort, Cleanliness and Convenience*. Berg, Oxford.

Shove, E., Watson, M., Hand, M. and Ingram, J. 2007. *The Design of Everyday Life*. Berg, Oxford and New York.

Shove, E., Chappells, H. and Lutzenhiser, L. (eds). 2010. *Comfort in a Lower Carbon Society*. Routledge, London.

Shove, E., Pantzer, M. and Watson, M. 2012. *The Dynamics of Social Practice: Everyday Life and How It Changes*. Sage, London.

Shove, E., Walker, G. and Brown, S. 2014. Material culture, room temperature and the social organisation of thermal energy. *Journal of Material Culture* 19(2): 113–124.

Smith, S. 2008. Owner-occupation: at home with a hybrid of money and materials. *Environment and Planning A* 40: 520–535.

Strengers, Y. 2010. *Conceptualising Everyday Practices: Composition, Reproduction and Change*. Working paper no.6, Centre for Design, RMIT University and University of South Australia.

Strengers, Y. 2011. Beyond demand management: co-managing energy and water practices with Australian households. *Policy studies* 32(1): 1–35.

Strengers, Y. and Maller, C. 2011. Integrating health, housing and energy policies: social practices of cooling. *Building Research and Information: The International Journal of Research, Development and Demonstration* 39(2): 154–168.

Part III
Housing/home and worlds of possibility

Andrew Gorman-Murray, *The External 5*, 2015.

Andrew Gorman-Murray, *The Internal 3*, 2015.

Andrew Gorman-Murray, *The Internal 7*, 2015.

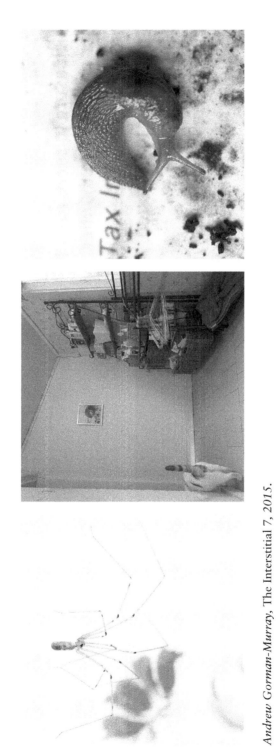

Andrew Gorman-Murray, The Interstitial 7, 2015.

10 Unbounding home ownership in Australia

Louise Crabtree

This chapter explores ruptures and incursions at the boundaries of 'home ownership' as currently upheld in dominant Australian discourse and practice, presenting two case studies that challenge normalised assumptions bound up in 'home ownership'. The chapter's core argument is that dominant discourses and practices of home ownership in Western societies are the product of the nexus of neoliberalism, individualism, colonialism and Judeo-Christianity. That nexus has created a constellation of economic, political, socio-cultural and material behaviours and expectations that conflate an idea of economic rationality and the search for ontological coherence, despite the frequently contradictory spatial and temporal imperatives of these two. Such contradictions are elided, and much legitimacy secured, through concomitant aesthetic, moral and spiritual registers of affect, such that 'privately owned' property as underpinned by globalised systems of debt is normalised to the extent that other arrangements are perceived either as inferior, or not at all. This renders invisible the actual characteristics and opportunities of housing lying 'outside' the dominant ownership discourse, as well as the contradictions within it. Therefore, the chapter presents case studies in which ontological coherence has been secured through property title other than 'ownership', and where dominant narratives and assumptions of ownership start to unravel.

The chapter proceeds as follows. It begins with a critique of the presumed citizen virtues of home ownership, drawing on work by Stephanie Stern. It then traces Stern's core arguments into the first case study: the current eviction of public housing tenants at the inner-city Sydney suburb of Millers Point. This case study demonstrates the inability of public policy to comprehend, recognise or prioritise the contribution of residents to place. This inability is then traced into the second case study: the promotion of home ownership on Aboriginal and Torres Strait Islander community lands. The chapter then reflects on these cases in the context of Australia's unsettled, uncanny condition, positing that the ruptures of ownership narratives might be expected from embracing property theory's relational turn in such a context.

W(h)ither the virtuous economic citizen?

The dominant Australian property narrative rests on an un/conscious assertion of owner occupation as a signifier of individual and financial success and worthiness (e.g. Darcy 2010). Private renting is generally portrayed as a poor cousin to owning or as a lifestyle choice for young, mobile households. Public rental housing is generally portrayed as housing of last resort or a transitional state on the pathway to more acceptable tenure forms (Atkinson and Jacobs 2008). That narrative incorporates two primary assumptions: owners' superior positioning and behaviour as citizens; and economic benefits to owners, whether through capital gains, withdrawing mortgage funds or borrowing against title. As Smith (2015, 62) argues, in such contexts ownership 'remains the preferred housing outcome of the responsible, risk-averse citizen'. Hence popular and policy discourses promote individual wealth creation through the individualised ownership of property, in which owners are lauded as hardworking, rate-paying[1] citizens (this will be explored below).

To unpack those dual assumptions and in the absence of comparable Australian research, there is much that can be gleaned from property law scholarship in the United States, a jurisdiction sharing Australia's preferential treatment and perception of ownership. Stephanie Stern (2009, 2011) analyses extant research showing that tenure duration and stability, rather than form, underpin residents' engagement in civic behaviours. Hence, '[w]here owners do make greater social contributions, length of residence, rather than ownership per se, appears to mediate many of the effects' and that 'it is of course possible, and even probable, that community, culture, or other social forces are more powerful prisms for local civic behaviour than ownership' (Stern 2011, 892, 893).

Stern (2011) shows that for owners and renters remaining in place for a similar duration, the primary difference in their civic behaviour is that owners will tend to join organisations at the sub-local level, such as neighbourhood associations – possibly as such organisations focus more directly on issues impacting property value – while renters are more likely to join broader-scale civic organisations. Stern (2011, 907) highlights that owners' concerns for their investment in housing 'drive not only socially beneficial investment, but also exclusionary behaviour, discriminatory practices, and other ill effects'. Stern argues that renters actually have a heightened sense of, and commitment to, social issues and organisation at a scale beyond their own property interest. That is, they may in fact be *more* civically minded than their owner counterparts – a possibility that jars with much dominant representation of renters as shiftless and not contributing to society as owners do (see Cook *et al.* 2013).

Stern highlights broader societal factors bound up in and impacted by tenure, including the widely touted psychological benefits of ownership, stating 'psychology research illustrates the primacy of social relations, not possessions, to self and flourishing' (2009, 1093). She argues that, consequently, 'misplaced belief in the psychological primacy of the home has encouraged the overproduction of home-protective legislation and added a gloss of moral legitimacy to rent seeking' (2009, 1093). Stern comes to the conclusion that it is the duration and stability of tenure, regardless of its form, that underpins and enables ongoing civic behaviour, and that underpin a coherent sense of self and one's place in the world. The one exception Stern highlights is the category of ritual or sacred objects, stating: 'If a specific, irreplaceable object is required for people to engage in important religious or cultural practices, then loss of that property is a loss of culture or spiritual belief' (2009, fn.11). Hence ontological coherence appears framed, and better addressed, through social rather than legalistic aspects of tenure – that is, through its purpose rather than its form. However, the materiality of the object of property also impacts ontological coherence, through its interpretation and manifestation as a socio-cultural artefact through ritual, repetition and memory. Iris Marion Young's (1997) work on ritual, repetition and memory in the articulation of 'home' is evocative here. The materiality of property (or specifically land or country) as a sacred or ritual object has particular resonance for considering the promotion of home ownership on Aboriginal and Torres Strait Islander lands, a point to which the chapter returns.

Such entanglements of affect, relations and practices make it difficult for property to function in a neoliberal market sense, as this requires a neatly defined object and a certain rapidity and ease of trade. Hence Blomley (2014) states that a phenomenon as inherently complex as property can only work through the assertion of prescribed boundaries – a process he calls 'bracketing'. Hence,

> Property as a bundle of relations is inherently productive of, and produced by, dense networks of power, membership, identity, connection, information and so on…. Yet for property to work, cuts in these networks must be made. Sharp lines have to be drawn around the subject and object of property, in order to ensure predictability.
>
> (Blomley 2014, 169)

This issue of predictability will also be returned to. At this point, it is the role of bracketing in dismissing non-owners as valid citizens that is of interest. The dominant Australian discursive bias towards ownership means that the bracketing of property creates or reinforces a substantial blindness to phenomena such as the work of rental occupants in adding value to not only the home in which they reside, but also to their broader neighbourhood and to society (as per Stern 2011). Such work and resultant improvements in amenity and social capital add tradable value to the home, generating the risk to the tenant of increased rent, or of an increased owner incentive to sell. Such work by tenants is not logical according to economic theories of action as it triggers conflicting signals to the tenant who derives use value from activities such as involvement in local schools and community groups, but faces the consequent risk of their own displacement. Consequently, Stern (2011, 921) argues, 'it is apparent how many citizenship behaviors may appeal to intrinsic sociability drives', which can be seen to override the tenant's 'rational' 'economic' interest. To further challenge the assumed supremacy of owners' civic participation, Stern raises the possibility that when under financial stress, owners may, not surprisingly, contribute *less* to civic life than stable tenants (2011, 930). It would therefore seem that in countries promoting ownership as the primary form of tenure on the basis of its presumed psychological and civic benefits, this is built on – yet seemingly oblivious to – the fact that the stability underpinning these has been primarily delivered through ownership due to precisely such preferential perception and consequently biased policy. That is, the positive feedback loop between policy and perception leads to a discursive conflation of the effects of the preferential treatment of a single tenure form, with the form of tenure itself.

The conflation of policy and perception reifies and exalts the presumed supremacy of ownership and its cohorts, regardless of whether this superiority is in evidence. That can manifest as blindness to policy bias towards ownership and blindness to the risks of 'ownership', interpreting failure of the tenure form as failure of the resident. Moreover, the conflation erases the ability of policy or public discourse to recognise the validity, agency or contribution of individuals not fitting within the boundaries of ownership. This suite of assumptions translates into insidious policy positions that can be read as moralistic and patronising arbitration of whom is to be housed, often as determined by relatively well-housed and stably-employed bureaucrats. This creates both the well-documented cycle of ongoing residualisation of the social housing sector alongside marginalising or demonising of its residents (Atkinson and Jacobs 2008) and the promotion of ownership as an economic salve and signifier of individualist success. To unravel these two, the chapter now turns to the case study of the current eviction of public housing tenants from social housing in Millers Point in Sydney, New South Wales.

Heritage and housing at Millers Point

> That's how we were always taught – to look after the place, look after the people that were here…. My family's paid over a hundred years' rent here…. If that's not paying your way, I don't know what is. Are we not good enough for the area now?… Losing the people that's lived here all their lives, it's just going to change the place completely.
>
> (Robert Flood quoted in *Sydney Morning Herald* 2014, np)

This quote is from a resident facing eviction due to the current sale of public housing in the prime harbourside suburb of Millers Point, Sydney. Much of this housing was constructed as housing for workers in dockside industries. Over time it was transferred through a number of government agencies to its current position within the New South Wales public housing authority's portfolio. As such, the former worker housing – some occupied across multiple generations of the same family, as indicated by resident Robert Flood, above – does not fit within the contemporary policy interpretation of public housing. Many of the properties house direct descendants of the original dockworker occupants – indeed, in its heritage listing the state asserts current tenants as part of the housing's social significance (Forced Out 2015).

The government announcement of its intention to evict residents and sell the properties at market rates triggered intense public media and online debate, and protest by residents and a broad range of supporters (Darcy and Rogers 2015). The proposed (and ongoing) sale concerns the heritage-listed worker houses and the Sirius Building – a purpose-built late-1970s neo-brutalist public housing complex of one- to four-bedroom units interspersed with rooftop gardens, occupied largely by elderly long-term residents. Darcy and Rogers (2015, 4) show how the state used a

> central discursive strategy, which constructed a narrative of public tenancy in Millers Point as 'unfair', and individual tenants as the beneficiaries of enormous and inequitable largesse delivered at the expense of genuinely disadvantaged existing and potential tenants in other neighbourhoods.

That discourse pits public tenants against each other on the basis of imputed subsidies, arguing that housing public housing tenants in more expensive urban locations represents unjustifiable tax expenditure. The properties' maintenance costs were also cited as a reason for the sale, while eliding instances where such cost was due to years of neglect by the state (Darcy and Rogers 2015). Hence, the discourse 'served simultaneously (and somewhat contradictorily) to devalue the houses as public assets while at the same time emphasising their increasing private value so as to ready them for private sale' (Darcy and Rogers 2015, 5).

Public and media responses to the announcement of the sale have been strong, with intense division over issues such as rights to the city, whether public housing should only ever be short-term, and whether accessible housing is a privilege or a right (see also Bell 2011). A stroll through the comments sections of internet media articles echo the state discourse, asserting that lower-income households should not expect to live in higher-value areas of the city, or have categorically no right to expect multi-generational affordability. While not representative, such commentary shows an assertion of a claim to space by propertied and working individuals, alongside resentful de-legitimisation of others presumed to not be manifesting the same citizen ideals. Hence, the ongoing financialisation of housing translates into moralistic assertions of who deserves to live where, why and on what terms, in line with their actual or perceived performance of a modern economic subjectivity. This deems that civic virtue and citizen rights be allocated according to the performance of identifiable and accepted forms of work.

Consequently, while recognised by the state as of heritage significance, the contribution and connection of Millers Point households to their suburb over decades are not able to counter, withstand or take credit for the increases in property value resulting from enhanced amenity and favourable perception of the suburb by entities overtly oriented to global economism (Darcy and Rogers 2015). The role of residents in constructing and retaining the now-desirable 'village feel' of the area are absent from the state discourse of asset management. Further, the state's role in streamlining and subsidising the suburb's development, most notably through developer concessions at adjacent Barangaroo, is largely absent from public debate. Millers Point illustrates a simplistic policy and media discursive orientation towards private market imperatives, whereby the state becomes asset disposal overseer and citizens become would-be bidders at widely reported auctions ending with sales figures in the millions of dollars. In addition to eliding and excising resident contributions, this orientation denies the role of the State in framing and enabling the market, and denies any alternate reading of residents' labour or of the deployment of state agency. It also denies the social structures and relationships underpinning public ownership, and any worth or legitimacy of public assets beyond their perceived exchange value, leading one academic to assert that '[s]ome in government seem to think that beautiful buildings on prime public land seem to be somehow wasted on us citizens, we who are the actual owners' (Thalis 2015, np).

As sales continue (with one property already having been re-sold by its initial buyer, who made $590,000 on the re-sale), the properties' market values may be impacted by the actions of the local government authority, which has control over renovation approvals. The local government authority has been hostile to the evictions and sales, and is restricting the extent of renovations on the sold houses, citing physical heritage objections (Hasham 2014). While recognised at the state level, the socio-cultural heritage bound up in the properties was not perceived to be of sufficient value to prevent the sale policy and process; however, once commodified, the properties' physical heritage gave the local government ammunition to enact its political agenda. In addition to illustrating policy elision of public housing tenants' contributions to place or rights to stability, the tenure enactment at Millers Point reveals the nonsense of buyers claiming private ownership. Given the history and ongoing contestation of the site, including the 1970s preservation of the housing stock by the pivotal industrial and community action of the Green Bans (Cook, this collection; Iveson, 2014), to claim the homes are privately owned seems frankly absurd. Certainly it highlights an instance wherein asserting a discourse of ownership as secured on the back of individuated success on the part of a singular, valorised citizen has to substantially deny history and context (Plumwood, 1995).

Home ownership on Aboriginal and Torres Strait Islander lands

The circumvention of residential history and context is perhaps nowhere more evident than in the recent public policy promotion of home ownership on Aboriginal and Torres Strait Islander lands, wherein the assertion of home ownership as economic salvation and disciplining device is clear. According to a federal Department of Families, Housing, Community Services and Indigenous Affairs (2010, 18) paper presenting the case for home ownership, the departmental concern was: 'How can Government achieve the right balance between facilitating home ownership for Indigenous Australians as an economic opportunity and supporting home ownership as a means to help build individual and social responsibility?'

In one question the federal government both offers economic hope – whether real or false – and implies extant irresponsibility among Aboriginal and Torres Strait Islander communities, displaying a similar blindness to residents' actual contributions as the Millers Point case. Tenure reform is posited as addressing both concerns, while eliding the existing complex socio-economic realities of Aboriginal and Torres Strait Islander lands and living arrangements, one example of which is explored below. Moreover, the Department's question draws upon an assumption that ownership drives a suite of acceptable citizen behaviours. The treatment of legitimacy is fairly brazen, as shown in a report promoting the subdivision and privatisation of community lands, based on the handful of instances where this had occurred:

> I have deliberately chosen not to include the names of the residential participants, as all are private people and buying a home is a very personal undertaking.... Each and everyone [*sic*] is a special, hardworking Territorian with legitimate aspirations who is helping to break limiting stereotypes.
>
> (Fagan 2012, 3)

Despite the socio-cultural complexity of tenure, and of the community governance contexts within which private ownership was being promoted, Fagan (2012) asserts individualised home ownership as a 'personal undertaking'. Moreover, legitimacy is ascribed to individuals as citizens of the Territory, and on the basis of an identifiably Western work ethic – 'special, hardworking Territorian with legitimate aspirations' (Fagan 2012, 3) – adding the offensive implication that extant praxis (which remains unspecified) is a limiting stereotype. For most communities, extant praxis is housing that was previously under community control, but which is currently under government control; similarly, many current community employment profiles would reflect the federal cancellation of Community Development Employment Program (CDEP) schemes, a move critiqued for increasing unemployment, undermining community services, and denying the program's ability to balance cultural obligations and 'work'. Altman (2007, 1–2) refers to the cancellation of CDEP as the single most destructive decision in Indigenous affairs policy that I have witnessed in 30 years of research and involvement in Aboriginal communities ... in many small communities remote from labour markets and commercial opportunities, CDEP participation is the only source of employment and income.... There is clear evidence that the flexibility of the CDEP scheme has accorded with Indigenous aspirations in many situations.

Hence, the unspecified 'limiting stereotypes' asserted as being transcended through the pursuit of ownership may in fact be referring to impacts of housing mismanagement and unemployment resulting from decades of multi-level policy interventions (Lea, 2012; Lea and Pholeros 2010). For many communities, land tenure is a nexus of community and cultural obligation, responsibility and reciprocity. Fagan (2012) is referring above to the few individuals interested in buying their home on community land, but such interest can represent a backlash against government policy intrusion into formerly community-controlled systems, in the absence of the capacity to re-establish such systems.

An example can be found in research undertaken among the Alice Springs Town Camp communities by the Tangentyere Council Research Hub as part of an AHURI project on community land trusts (Crabtree *et al.* 2015), which highlighted the almost nonsensical proposition of home ownership being put to Camp residents by government representatives. The Town Camps communities hold title to their land and housing through perpetual leases from the Territory government, governing and allocating these according to community-based protocols reflecting complex histories of multi-national, tribal, kinship and community obligations. This includes suspension of Native Title over the Camps by the Traditional Owners, in acknowledgement of the Camps' histories as places of refuge when Aboriginal and Torres Strait Islander individuals from several nations were excluded from Alice Springs during the nineteenth and twentieth centuries.

Perhaps not surprisingly, Town Camp communities feel a strong sense of ownership, particularly given some Campers are also Traditional Owners, with ties to Country that extend for millennia. In this context, to approach the head tenant of a Town Camp household and offer ownership seems an exercise in patronising redundancy. Indeed, one interviewee stated that many Campers question 'why should we want to ... go for private home ownership when we already own it?' (Crabtree *et al.* 2015, 71). Regarding home ownership on community lands, one public servant stated (emphasis added):

I think the reasons were actually not primarily economic although you know they're rational people, they're not going to make what they regard as an uneconomic decision ... but more to do with having an asset that is theirs that will remain in their families and that they can pass on to their children, *to get out of the public housing administration, to have a greater sense essentially of control over their lives* by owning their own home.

(Anonymous interviewee)

Hence, interest in home ownership was at least partly a response to changes to community governance enacted by the federal government through the suite of policies known as the Intervention,[2] and by changes to local governance as enacted by the Northern Territory government. So, the (albeit minimal) interest was largely the result of public policy impacts on community governance and autonomy, such that private ownership was being considered as a way to escape such interference. Hence, as being proposed for communities and as with the Millers Point sell-off, ownership was nothing like a private affair and was rife with government involvement, including: subdividing community lands; revoking multiple layers of community and household leases enacted under the Intervention; drafting and enacting new individual leases to sit outside of the Territory's Residential Tenancies Act; and enabling mortgages through support and regulation of the specialist lender Indigenous Business Australia. So while being proposed as 'a personal undertaking', ownership represents a deeply entangled affair, shot through with contradictory imperatives. For example, one public servant stated:

we did a survey on at what level they considered housing to be in their way of life and it was about number three or four. The first one was family ... housing was down there so it's, in some cases you sort of think oh yeah, housing should be top of the list but it's not always looked at that way.

(Anonymous interviewee)

Apart from highlighting the primacy of family, the quote is illustrative for highlighting that the public servant did not think family should be the priority. Perhaps not surprisingly, given such disparate assumptions, communities remain suspicious of policy directives, including the recent promotion of home ownership. Another interviewee spoke frankly about the history of churn and novelty in policy, and resultant suspicion among communities:

if you look at the history of Indigenous affairs that would probably make you a screaming conservative because so many experiments have bitten the dust, you know, government policy's changed, crucial people have died, resigned.... I completely understand why Traditional Owners are very cautious about any 'You Beaut' ideas from government.

(Anonymous interviewee)

Given the contradictions and inconsistencies between policy imperatives and community objectives, then, it seems that what is occurring is an attempt to replace a current use of Western property law in ways that reflect and respect community aspirations with one that is more familiar and acceptable to policy and financial entities – the individually titled, indebted home occupied by an individual labouring in recognisable and sanctioned ways. This resonates with Wolford's (2007) assessment that beneficiaries of land tenure reform must be able to appropriately demonstrate their productivity. The churn and inconsistencies highlighted above also resonate with Lea's (2012) exploration of the generation of chaos and consequent self-justifying policy freneticism by bureaucrats in this space.

Moreover, the research with the Town Camps found that, should they be desired, core economic arrangements of mortgage-backed ownership could be articulated through community-governed subleases on the extant perpetual leases, requiring nether the excision of individual lots to freehold, nor the presence of government intermediaries (Crabtree *et al.* 2015). Many residents expressed concern regarding the implications for the relationship between a householder, their Town Camp association or corporation and the overarching community governance body, should a home lot be excised and made freehold, given that current titling and governance systems express and uphold core community relationships. This also raised the issue of whether the suspended Native Title would be triggered by any spot deployments of freehold.

The Alice Springs Town Camps reflect a situation where the social relationships reflected and enacted through property are the foremost concern of the residents with regard to tenure, yet are absent in public policy. Camp tenure is deeply and overtly relational, and the form of its relationality jars with that of dominant models of 'ownership'. Given the central role of Country in cultural and spiritual praxis, for Traditional Owners the Camps might also represent an example of Stern's singular exception of the sacred object. Here, the tenure and corresponding governance arrangements are themselves manifestations of precisely the types of 'social interactions and ties' that 'are the bedrocks of psychological thriving' (Stern 2011, 1110). This is an important extension of Stern's argument: while she discusses the role of particular owned objects' sacredness in proving their exceptional nature, her discussion carries an imputed delineation between the object's sacredness and the system through which it is owned. Whereas the complex patterns of Aboriginal and Torres Strait Islander relationships to each other and to place as manifest in and through Country present an example in which both the object and the system through which it is 'owned' comprise a social, cultural and spiritual nexus (Small and Sheehan 2008).[3] Consequently, the sacredness of the object inflects the corresponding arrangements of its ownership/governance. Indeed, in the framework of Country, such separation between the land and its governance is untenable and nonsensical. Considering the ramifications of Country highlights the elision of historical, and ongoing, Judeo-Christian assumptions in attempts to assert dominant models of property (Crabtree 2013; Rose 2004).

Promoting home ownership for Aboriginal and Torres Strait Islanders as a wealth creation strategy is similarly blind to context. For many such properties, the pool of prospective future buyers is small, and presuming wealth creation through tenure reform and capital gain ignores the array of broader mechanisms that enable such gains. If the pool of buyers is limited, capital gains cannot be realised unless access is increased. This is a critical issue that some communities have been hurt by: respondents in earlier research referred to instances where Aboriginal community housing agencies in NSW had made some of their rental stock available for purchase by tenants at sub-market rates (Crabtree *et al.* 2012). In many instances, the buyers had then sold to real estate agents who had taken advantage of residents' lack of both market knowledge and ownership experience to offer relatively inflated (but still sub-market) prices. The agents had then sold the properties at greatly higher market rates, reaping windfalls. Properties were thus lost from the Aboriginal estate, and in many instances, due to precarious employment and gains having been spent, residents had wound up on waiting lists for a now reduced pool of community rental houses.

For many such communities, employment and appropriate economic development are primary issues, so 'the real impediments to home ownership are not communal title or the Land Councils, but factors that affect most Australians: the cost of houses and the ability of people to pay for them' (Ross 2013, np). This speaks directly to the contextual factors that enable debt-financed ownership but which are not considered in its promotion – a situation in which debt-financed ownership becomes a very precarious proposition. In the context of communities without a Western economic base, deployment of private ownership can act to erode both the cultural bonds upon which psychological wellbeing actually rests (Stern 2011), and extant tenure stability. That is, it can lead to both compromise of cultural and individual wellbeing, and loss of title – a potential outcome that in the context of Aboriginal lands containing substantial mineral or fossil fuel deposits, must be viewed with at least some suspicion.

In such contexts, rather than the pinnacle of success and security, mortgagee ownership can therefore be highly inappropriate and risky, and translate into loss of home, economic security, community stability and ontological coherence (Hulse *et al.* 2010). This is highlighted by Smith (2015) with regard to the inherent contradiction enacted in borrowing against the home, whereby ownership's ostensibly secure nature is fundamentally compromised through leverage. The AHURI research found that the potential for debt-financed ownership to lead to loss of title is significant enough for lenders other than the sector-specific lender, Indigenous Business Australia, to shy away. Given the inherent yet hidden instability of borrowing to buy, Stern (2011, 929) states that ownership as a citizen virtue is 'at best, a theory for good economic times', illustrating that the financialisation of housing means that in sour economic times, not only the economic but also the social benefits of ownership are compromised.

Hence dominant narratives focusing on ownership's assumed economic benefit and associated citizen behaviours are very partial stories obscuring the actual factors underpinning or constraining citizen behaviours and economic activity, and focusing on select forms of these – namely, those deemed acceptable by the state or its representatives. In Australia, this refers to citizens articulating an individualised, exclusionary relationship to a singular, static location (Prout and Howitt 2009) and engaging in recognisable industrial capitalist forms of economic activity (Wolford 2007). This not only creates phenomena such as the discourse surrounding home ownership on Aboriginal and Torres Strait Islander lands; as seen in both case studies, it negates substantial aspects of the social, economic and cultural relationships that create and are affected by property.

Property, housing, decoloniality

The above case studies represent instances when the dominant narrative of home ownership as private, individualised, superior to renting and an avenue for wealth creation start to unravel, as do assertions of homeowners as the sole proprietors of civic virtue. The defence of the ownership discourse ramparts in the face of the contradictions explored above, and by others such as Smith and Stern, begs the question – why is the narrative upheld despite such apparent contradictions and inconsistencies? To consider this in a settler colonial nation, it is worth exploring the state's need for an identified object of governance, and the implications of ongoing dialogue, tension and entanglement of diverse ontologies.

It is therefore worth returning to Scott's (1999) thesis regarding the state requiring legibility, and the work of Lea (2012) and Lea and Pholeros (2010) on interventionist bureaucratic busywork, alongside Blomley's (2014) reflections on bracketing. In this light, property can be seen as an act of simplification enabling bureaucratic apparatuses to function, and upholding particular understandings of humanity's place in the world. In Western industrialised societies, these twin, entwined aspects are steeped in the trappings of Judeo-Christian ontologies (Crabtree 2013; Rose 2004) and in the epistemologies and processes of coloniality (Graham 2011). Examining housing policy impacts in Australia reveals the act of creating legibility (or bracketing) as an act of policing the borders of the type of citizen that is deemed acceptable, and of affiliated sanctioned economic behaviours (also Wolford 2007). The role of property and of state regulation vis-à-vis housing regulation, provision and taxation, reflects and reinforces particular suites of relationships. In a settler colonial nation such as Australia, this has to be reviewed in light of the social, political and economic imperatives of Empire, and in light of engagements between Aboriginal and Torres Strait Islander and colonial experiences of, and aspirations for, place.

Numerous authors refer to entanglements of diverse ontologies and lived experiences in Australia, and a degree of – albeit uneven – bi-directional adoption and adaptation (e.g. Abramson 2000; Dolin 2014; Gelder and Jacobs 1998; Jacobs 1996). In Australia, while a singular colonial legal system appears to dominate (and is generally able to), myriad systems of law are actually in dialogue. Strang (2000, 94) refers to a consequent 'convergence of Aboriginal and European values relating to land' while a growing body of Australian legal geography scholarship grapples with people–place relations as manifest through law in a de/colonial context (e.g. Bartel *et al.* 2013). That scholarship is part of a broader relational turn in property theory, spanning law and legal geography, and bringing to the fore the fundamental construction and enactment of property law as a codified and situated system of interpersonal and environmental relationships. Consequently, in settler colonial societies, ongoing articulations of property law must be cognisant of the history and persistence of dispossession (Moreton-Robinson 2015), and open to the 'uncanny' ambivalence (Gelder and Jacobs 1998) triggered by ongoing Aboriginal and Torres Strait Islander voices, whereby something as ostensibly familiar as 'ownership' is rendered incoherent to a dominant Western reading. Such opening or unbounding of assumptions hopefully can do transformative work (see Keenan 2010); certainly, considering Country in the city emerges as a pressing issue. However, this is in an environment of diminished policy capacity and imagination (Gurran and Phibbs 2015; Jacobs 2015), if not the ongoing bureaucratic creation of chaos (Lea 2012).

In contrast to modernist assertions of property as a tradeable commodity, in Australia, Country is upheld as an understanding of place that is profoundly enmeshed in ideas of the sacred in immediate, visceral ways that directly implicate human responsibility and agency. This presents a mirror in which colonial enactments of property are made to take a good, hard look at themselves – not with some naive, nostalgic sense of reverence to the ostensibly unique presence of the sacred in such understandings (Gelder and Jacobs 1998), but with an eye to the hidden work of colonial property enactments, and in an atmosphere of unsettlement. Gelder and Jacobs (1998, 20) refer to the unsettlement caused by the sacred in this context: 'its often massive legal costs, its demands on land, its apparent disregard for capitalist interests and the welfare of the nation, its reminder that Aboriginal people have ongoing agendas', asserting Australia's 'uncanny' condition as a productive space constructed by tensions between reconciliation and its apparent impossibility. Since that writing, Australia's uncanniness has perhaps intensified, with the Apology sitting alongside the Intervention,[4] and an ongoing incapacity of policy or ideology to process the fact that the majority of Australia's first peoples live in urban areas (Prout 2011), adopting, adapting and/or transgressing dominant narratives according to their needs and aspirations.

This chapter's case studies highlight property's relationality – the social, cultural, political, affective and spiritual dimensions that are simultaneously codified and obscured in dominant property articulations. The collapse of the dominant ownership narrative in the case studies reveals the role of the narrative in policing the borders of acceptable citizen forms and behaviours, regardless of the accuracy or fallacy of such delineation. In this way, tenure discourses can be seen as deeply embedded in political acts of creating (or, perhaps, increasingly struggling to maintain) particular public imaginaries and identities. In unbounding ownership, and especially in light of its relationality, it is possible to imagine property articulations that speak to a broader, dynamic range of socio-economic configurations. This is not to assert that property will actually do such work, or that public and political imaginaries will respond well or creatively. It is, however, to lay the challenges bare. Moreover, in reference to Gelder and Jacobs' (1998) productive ambivalence, the intention here is not to posit 'solutions', but to step (out) into that ambiguous space.

Notes

1 'Rates' being land taxes payable to local municipalities and used to fund the operations and obligations of municipalities, such as the provision of local roads, waste services and local planning oversight.
2 The Northern Territory National Emergency Response Act 2007 (Cth); Social Security and Other Legislation Amendment (Welfare Payment Reform) Act 2007 (Cth); Families, Community Services and Indigenous Affairs and Other Legislation Amendment (Northern Territory National Emergency Response and Other Measures) Act 2007 (Cth); Appropriation (Northern Territory National Emergency Response) Act (No. 1) 2007–2008 (2007) (Cth); Appropriation (Northern Territory National Emergency Response) Act (No. 2) 2007–2008 (2007) (Cth); *Stronger Futures in the Northern Territory Act 2012* (Cth); Stronger Futures in the Northern Territory (Consequential and Transitional Provisions) Act 2012 (Cth); and Social Security Legislation Amendment Act 2012 (Cth).
3 Not all Campers are Traditional Owners (TOs), and I am not suggesting the Camps are necessarily manifestations of Country, nor to infer that Campers who are not TOs would assert a claim to the Camps as their Country, nor do I claim to speak for (or of) the full intricacies of the Camps. To summarise, governance and residence on the Camps represent a complex, nuanced arrangement between TOs and Campers whose Country is further afield; this includes the suspension of Native Title over the Camps by the TOs in recognition of the Camps' histories as a place where diverse communities could be safe and strong in the face of displacement from their Country and/or from the town of Alice Springs.
4 The Apology to Australia's Indigenous Peoples ('the Apology') was hand-scripted and delivered as a speech by the incumbent Prime Minister the Hon. Kevin Rudd in 2008 as one of the first actions of the newly elected government when Parliament resumed. Many household, school, community and public events were held to watch the live broadcast of the Apology, including the gathering of several thousand people on the lawn of Parliament House in front of large screens and speakers. Despite the emotion and significance of the Apology, many feel it has changed little in terms of actual outcomes for Aboriginal and Torres Strait Islander peoples, and some see it as having been largely 'for show' (ABC News for Australia Network 2014).

References

ABC News for Australia Network. 2014. *The Stolen Generation Apology: Has Anything Changed?* Retrieved 11 January 2016 from www.youtube.com/watch?v=KM8l--Il8gs

Abramson, A. 2000. Mythical land, legal boundaries: wondering about landscape and other tracts. In A. Abramson and D. Theodossopoulos (eds) *Land, Law and Environment: Mythical Land, Legal Boundaries.* Pluto Press, London.

Altman, J. 2007. *Neo-Paternalism and the Destruction of CDEP*, Centre for Aboriginal Economic Policy Research Topical Issue No. 14/2007. Retrieved 11 January 2016 from http://caepr.anu.edu.au/sites/default/files/Publications/topical/Altman_Paternalism.pdf

Atkinson, R. and Jacobs, K. 2008. *Public Housing in Australia: Stigma, Home and Opportunity*, Paper No. 01 Housing and Community Research Unit. Retrieved 11 January 2016 from http://eprints.utas.edu.au/6575/1/public_housingLR.pdf

Bartel, R., Graham, N., Jackson, S., Prior, J.H., Robinson, D.F., Sherval, M. and Williams, S. 2013. Legal geography: an Australian perspective. *Geographical Research* 51(4): 339–353.

Bell, K. 2013. Protecting public housing tenants in Australia from forced eviction: the fundamental importance of the human right to adequate housing and home. *Monash University Law Review* 39(1): 1–37.

Blomley, N. 2014. The ties that blind: making fee simple in the British Columbia treaty process. *Transactions of the Institute of British Geographers* 40(2): 168–179.

Cook, N., Taylor, E. and Hurley, J. 2013. At home with strategic planning: reconciling resident attachments to home with policies of residential densification. *Australian Planner* 50(2): 130–137.

Crabtree, L. 2013. Decolonising property: exploring ethics, land, and time, through housing interventions in contemporary Australia. *Environment and Planning D: Society and Space* 31(1): 99–115.

Crabtree, L., Blunden, H., Milligan, V., Phibbs, P., Sappideen, C. and Moore, N. 2012. *Community Land Trusts and Indigenous Housing Options*, Final Report No. 185. Australian Housing and Urban Research Institute, Melbourne.

Crabtree, L., Moore, N., Phibbs, P., Blunden, H. and Sappideen, C. 2015. *Community Land Trusts and Indigenous Communities: From Strategies to Outcomes*, Final Report No. 239. Australian Housing and Urban Research Institute, Melbourne.

Darcy, M. (2010) De-concentration of disadvantage and mixed income housing: a critical discourse approach. *Housing, Theory and Society* 27(1): 1–22.

Darcy, M. and Rogers, D. (2015). Place, political culture and post-Green Ban resistance: public housing in Millers Point, Sydney. *Cities.* http://dx.doi.org/10.1016/j.cities.2015.09.008

Department of Families, Housing, Community Services and Indigenous Affairs (FaHCSIA). 2010. *Indigenous Home Ownership Issues Paper.* FaHCSIA, Canberra.

Dolin, K. 2014. Place and property in post-Mabo fiction by Dorothy Hewett, Alex Miller and Andrew McGahan. *Journal of the Association for the Study of Australian Literature* 14(3): 1–12.

Fagan, M. 2012. *Towards Privately Financed Property on Indigenous Land in the Northern Territory.* Northern Territory Government, Darwin.

Forced Out. 2015. *Preview: Forced Out – The Documentary* [video]. Retrieved 11 January 2016 from www.youtube.com/watch?v=KsKkBId2_gQ

Gelder, K. and Jacobs, J.M. 1998. *Uncanny Australia: Sacredness and Identity in Postcolonial Australia.* Melbourne University Press, Melbourne.

Graham, N. 2011. *Lawscape: Property, Environment, Law.* Routledge, New York.

Gurran, N. and Phibbs, P. 2015. Are governments really interested in fixing the housing problem? Policy capture and busy work in Australia. *Housing Studies,* August: 1–19.

Hasham, N. 2014. No fancy extensions, City of Sydney tells new Millers Point buyers, *Sydney Morning Herald*, 27 October. Retrieved 11 January 2016 from www.smh.com.au/nsw/no-fancy-extensions-city-of-sydney-tells-new-millers-point-buyers-20141026-11a00v.html

Hulse, K., Burke, T., Ralston, L. and Stone, W. 2010. *The Benefits and Risks of Home Ownership for Low–Moderate Income Households*. Australian Housing and Urban Research Institute, Melbourne.

Iveson, K. 2014. Building a city for 'the people': the politics of alliance-building in the Sydney Green Ban movement. *Antipode* 46(4): 992–1013.

Jacobs, J.M. 1996. *Edge of Empire: Postcolonialism and the City*. Routledge, London.

Jacobs, K. 2015. The 'politics' of Australian housing: the role of lobbyists and their influence in shaping policy. *Housing Studies*. DOI: 10.1080/02673037.2014.1000833

Keenan, S. 2010. Subversive property: reshaping malleable spaces of belonging. *Social & Legal Studies* 19(4): 423–439.

Lea, T. 2012. When looking for anarchy, look to the state: fantasies of regulation in forcing disorder within the Australian Indigenous estate. *Critique of Anthropology* 32(2): 109–124.

Lea, T. and Pholeros, P. 2010. This is not a pipe: the treacheries of Indigenous housing. *Public Culture* 22(1): 187–209.

Moreton-Robinson, A. (2015) *The White Possessive: Property, Power, and Indigenous Sovereignty*. University of Minnesota Press, Minneapolis.

Plumwood, V. (1995). Feminism, privacy and radical democracy. *Anarchist Studies* 3, 97–120.

Prout, S. 2011. 'Urban myths: exploring the unsettling nature of Aboriginal presence in and through a regional Australian town'. *Urban Policy and Research* 29(3): 275–291.

Prout, S. and Howitt, R. 2009. Frontier imaginings and subversive Indigenous spatialities. *Journal of Rural Studies* 25(4): 396–403.

Rose, D.B. 2004. *Reports from a Wild Country: Ethics for Decolonisation*. University of New South Wales Press, Sydney.

Ross, D. 2013. *Communal Title no Obstacle for Aboriginal Home Ownership*, Central Land Council media release. Retrieved 11 January 2016 from www.clc.org.au/media-releases/article/opinion-piece-communal-title-no-obstacle-for-aboriginal-home-ownership.

Scott, J.C. 1999. *Seeing Like a State: How Certain Schemes to Improve the Human Condition Have Failed*. Yale University Press, New York.

Small, G. and Sheehan, J. 2008. The metaphysics of indigenous ownership: why indigenous ownership is incomparable to western concepts of property value. In Simons, R.A. Malgren, R., and Small, G. (eds), *Indigenous Peoples and Real Estate Valuation*. Springer, Netherlands.

Smith, S.J. 2015. Owner occupation: at home in a spatial, financial paradox. *International Journal of Housing Policy* 15(1): 61–83.

Stern, S.M. 2009. Residential protectionism and the legal mythology of home. *Michigan Law Review* 107(7): 1093–1144.

Stern, S.M. 2011. Reassessing the citizen virtues of homeownership. *Columbia Law Review* 111(4): 890–938.

Strang, V. 2000. Not so black and white: the effects of Aboriginal law on Australian legislation. In A. Abramson and D. Theodossopoulos (eds) *Land, Law and Environment: Mythical Land, Legal Boundaries*. Pluto Press, London.

Sydney Morning Herald. 2014. Millers Point: a community under the hammer. Retrieved 11 January 2016 from www.smh.com.au/interactive/2014/millers-point/home.html

Thalis, P. 2015. Sydney risks becoming a dumb, disposable city for the rich. *The Conversation*, 2 March 2015. Retrieved 11 January 2016 from https://theconversation.com/sydney-risks-becoming-a-dumb-disposable-city-for-the-rich-38172

Wolford, W. 2007. Land reform in the time of neoliberalism: a many-splendored thing. *Antipode* 39(3): 550–570.

Young, I.M. 1997. *Intersecting Voices: Dilemmas of Gender, Philosophy and Policy*. Princeton University Press, Princeton.

11 Performing housing affordability

The case of Sydney's green bans

Nicole Cook

Introduction: the green bans revisited

This chapter explores the relationship between urban social movements, civic action and housing policy, taking Sydney's green bans from 1971 to 1974 as a case study. This period of urban activism saw the interruption of urban development plans by the New South Wales Builders Laborers Federation (NSWBLF), the union representing builders and labourers in the State of New South Wales (NSW) who, in support of resident activists, refused to work on building sites that involved the demolition of public housing, heritage buildings or the denuding of green spaces in diverse neighbourhoods (Mundey 1981). With green bans placed on over 40 sites, the union's refusal saw delays in urban development on many high-profile projects, including in Sydney, the redevelopment of The Rocks, the Theatre Royal and Centennial Park (Burgmann and Burgmann 1998). Affordable housing and heritage buildings in inner-city precincts including Woolloomooloo and Glebe were also protected. Comprising a cross-class coalition of workers, residents and professionals that sought to put 'people before profits', the green bans epitomised grass-roots urban activism during a period of inner urban restructuring (Hardman and Manning 1975). This action also had lasting influence on urban renewal processes through emphasis on affordable housing, green space and heritage conservation, long before such values were articulated in formal planning legislation and urban policy.

The chapter takes as its focus the redevelopment of The Rocks, including the suburb of Millers Point located at the southern end of the Sydney Harbour Bridge, that from 1971 to 1974 was protected by a green ban. Located in the inner city near Circular Quay, the site is today part of a revitalised and gentrified tourist enclave, home to restaurants, bars, hotels, museums and arts and craft markets. Then, as today, the redevelopment of this site marked a tension between low-income residents, long living in the neighbourhood in affordable (social) housing and capital accumulation through urban redevelopment. While today many of these houses are subject to state-led privatisation (Hasham and Johnston 2014), the green bans effectively protected these and other threatened houses and homes for more than four decades. The event of the green bans has been interpreted as a class-based social movement (Birmingham 2000; Jacubowicz 1984), as an exemplar of community-based unionism (Burgmann and Burgmann 1998) and, by Anderson and Jacobs (1999) in their feminist re-reading, as a struggle over public and private space to which the union was strategically recruited by resident action groups. More recently, the cultivation and expression of shared interests across difference, as typified by the green bans, has been presented by Iveson (2014) as the basis for social justice in the city.

One of the most significant outcomes of this unlikely alliance in the case of The Rocks was to maintain the supply of affordable housing in waterfront urban renewal. In doing so, the green bans effectively unbound housing/ home from a regime of speculative office and commercial redevelopment and, through practices of striking, protest and occupation, moved social and ecological concerns centre-stage in urban waterfront redevelopment with significant implications for housing and planning policies in NSW in subsequent years. The steer from the green bans is that the material world – in this case the stilled building site – plays an important role in forming new configurations of dwelling, society and space.

In this chapter I build on existing accounts of the green bans to position the event as a social movement that manifested as an informal but powerful institution of urban policy, housing and planning which unsettled, (partly) reordered and established new policy frameworks for urban planning and housing in subsequent decades. I first set out the institutional challenges facing the green ban activists, flagging the formal impossibility of retaining affordability in urban redevelopment in terms of housing and planning policy at the time. From here, I explore the ways the green bans led to innovation in waterfront development with particular reference to the retention and expansion of the stock of affordable housing, resident rights to remain in place and innovation in planning processes. Following these two mostly descriptive sections, I next consider in three short theoretical sections: the implications of the green bans for conceptualisations of housing and home; the role of nonhuman collaborators in the achievement of the green bans; and the profoundly contingent character of this event. The green bans effected significant change in urban development, planning and housing in NSW, but these innovations did not emerge from the institutions and authorities that normally manage, develop and implement urban policy. To these ends the chapter also explores the virtuality of the green bans, where virtuality refers to 'functioning otherwise than its plan or blueprint' (Grosz 2001, 130). I conclude with reflections on the significance of the green bans for social movements and transformative policy innovation more broadly.

Contesting Millers Point: the 1970s

In August 2014, an auction for 119 Kent Street, Millers Point was held in a secret non-publicised sale (Hasham and Johnson 2014). The house was the first of 293 sales that were scheduled to follow in this inner Sydney suburb, a harbourside location that throughout the twentieth century was a settlement site for successive generations. While Millers Point has retained its colonial architecture, the processes of gentrification and urban renewal have placed pressure on the area's working-class heritage, housing and communities. The properties placed on the market had been in state ownership for most of the twentieth century, initially providing for low-income residents and managed by the Maritime Services Board, offering affordable housing at minimal rents. In this time, the rights of low-income residents to social housing in both Millers Point and the nearby Rocks precinct were challenged by five redevelopment proposals (Roddewig 1978, 22), including most notably the 1971 Rocks Redevelopment Plan by the Sydney Cove Redevelopment Authority (SCRA). Described by planning historian Robert Freestone (2000, 135) as an 'operatic ensemble of skyscrapers and superblocks', the 1971 plan proposed demolition of two-thirds of the site, including the removal of 416 residents and their homes.

In Pat Fiske's (1984) excellent documentary of the green bans, *Rocking the Foundations*, Nita McRae, the president of The Rocks Resident Action Group and long-term social housing tenant in Millers Point, provides an on-camera interview in front of the home from which she had been (unsuccessfully) evicted as part of the 1971 SCRA Plan. With no recourse to challenge the decision through procedural or political pathways, McRae recalls the efforts of the resident group to find support among the building unions. Just four months before, the NSWBLF had, in association with the Federated Engine Drivers & Firemen's Association, refused to clear a bushland site in preparation for new housing development in support of the middle-class resident action group at Kelly's Bush in the suburb of Hunters Hill. It was this initial refusal that, more than 18 months later, became known as the first green ban (Mundey 1981, 105). The name 'green ban' was an environmental variation on the more common 'black ban' that referred to building strikes over wages and conditions. The redevelopment of 53 acres of prime inner-city real estate in The Rocks was nonetheless a step up from the modest proposal for 57 townhouses at Kelly's Bush (Cook 2007, 96), and more immediately aligned with the working-class members of the union. When asked by Nita McRae if it was the policy of their union to destroy working-class homes, there was little resistance from members of the NSWBLF, who voted to withdraw their labour from the demolition of residents' homes and to remove (by force) any non-unionised labour to enable the working-class residents to negotiate with the planning authority.

Unlike planning processes at the end of the century that, for better or worse, had made a collaborative turn (Healey 1997), there was no recourse for residents to contest this redevelopment though procedural or political pathways in Sydney in the 1970s. Under the Local Government Act 1919 (NSW), adjacent property owners could express an opinion about development proposals but there was no opportunity or requirement to consider third-party objections from residents who did not own their property or who were not immediately impacted by a development. The informal processes of resident engagement that eventually found their way into the NSW planning system were still some decades away. In the case of The Rocks, the annexure of this jurisdiction from Local Government Planning authority through the creation of the SCRA further limited public input into processes of neighbourhood change, such that the new authority had the right to approve plans without public exhibition, to raise rents and evict residents.

There is no doubt that the process of urban renewal at The Rocks was part of a middle-class restructuring of space (Shaw 2008). The plans were designed to increase office space for large corporations and high-rise luxury hotels that, according to the 1968 Sydney Region Outline Plan (1970–2000) would contribute to Sydney's position as 'Australia's greatest city, of world status and importance' (quoted in Sandercock 1977, 187). Described by the Director of the SCRA as 'a rather depressing area and not really suitable as an entry to the city of Sydney' (Magee in Fiske 1984: 45 min), plans for renewal generated uncertainty among the existing population, many of whom would receive notices of eviction (Roddewig 1978). The assertion by the union on behalf of residents that the SCRA guarantee the provision of low-income dwellings as part of the urban renewal process was regarded by the Director of the SCRA as impossible (Burgmann and Burgmann 1998, 198).

While the experience of residents at The Rocks was framed by the authority of the SCRA, the trend to repossess and demolish housing was widespread for both office and freeway construction at the time. While for much of the twentieth century economic growth and urban form in Sydney were linked through suburbanisation, consumption and manufacturing employment, the urban restructuring of the early 1970s saw opposition to office and commercial high-rise development among inner-suburban communities. Daly (1982) identifies the factors shaping urban restructuring as a combination of foreign investment, the changing role of Sydney in an internationally competitive mining sector and its increasing financial significance. However, the process of urban change was facilitated by significant political intervention. In 1967, for instance, the Premier of the State sacked the entire Sydney City Council, redrew the electoral boundaries and installed a caretaker government to facilitate high-rise development (Sandercock 1977). Planning approvals doubled and areas previously outside the remit of the council were incorporated into its approval authority. While political and planning influence is expected where elected officials have planning power, the politicisation of the planning system in this time was associated with widespread use of eviction and termination, along with modification to residents' tenancy powers (Burgmann and Burgmann 1998). This shift facilitated high-rise development in spite of considerable neighbourhood impacts and opposition. In Sydney in the 1970s, then, urban renewal was tied to a highly politicised drive for capital accumulation in which resident attachments to place were secondary concerns.

Innovation in urban renewal: protecting affordable housing

Elizabeth Grosz (1999) has argued that innovation emerges in such contexts of impossibility. The 'new' is not merely the realisation of what is possible, but the bringing into the world of something that is unknown and unexpected. This is not a distant utopia, but something closer to Amin's (2012) 'pragmatic urbanism'; a metamorphosis of humans and materials that takes place in the world as it is. In Millers Point, the likelihood of planning modification that reflected resident viewpoints or incorporated affordable housing was approaching such an impossibility: in planning law and policy there were no rights for residents, no pathways of influence, and there was no chance that social housing tenants would be incorporated into the process of urban renewal. Still, experimentations in housing policy emerged through the effort of workers and residents whose livelihoods and dwellings were the subject of urban planning and development processes, but whose presence was also formally excluded from consideration in urban planning and development strategies.

At the time that The Rocks Resident Group sought help from the unions, the NSWBLF was explicitly seeking stronger connections with communities around urban change. The union was at the beginning of a new phase of democratisation and community engagement inspired by the leadership's commitment to New Left ideologies. The 1968 leadership team specifically advocated the transfer of power away from the executive to the members by bringing organisers' wages in-line with the workers', introducing limited tenure of office for officials and opening up meetings of the executive to all active members (Burgmann and Burgmann 1998). The NSWBLF also embarked on a concerted campaign to dislodge wage negotiation from arbitration, and with greater willingness to strike won substantial wage gains in the five-week margins strike in 1970 (Thomas 1973, 17). It was nonetheless isolated among the other building unions, particularly in its engagement with the politics of the New Left and its insistence that workers had a right to have a say over what they built. This philosophy of community-unionism was important at the time, and marks diversity within the building unions about the role of collective labour in supporting communities beyond the needs of the builders and workers themselves.

The SCRA plan was released in February 1971 and by November the union had enacted its support for both the residents and the heritage buildings in the area, asserting 'it would not move a single brick until the 416 residents forced to move had been satisfactorily rehoused' (Burgmann and Burgmann 1998, 196). In the first set of evictions of low-income residents on 14 January 1972, 85 residents were required by law to leave their home to enable the first phase of demolition ahead of plans for a luxury hotel and office block. In support of residents, the NSWBLF confirmed that it would maintain the ban until such time as the SCRA would meet with and guarantee that residents would be relocated within the area at the same rents. At the same time, the union only provided labour for historic building restoration (Burgmann and Burgmann 1998).

Footage captured of the 1973 protest march at The Rocks testifies to the union's commitment to affordable housing policy and the rights of low-income residents. In it, Jack Mundey – secretary of the NSWBLF – addresses a crowd of workers and residents who supported the retention of social housing within The Rocks. In this address he outlines a vision of urban renewal that deviated significantly from the public policy at the time and from traditional union concerns, stating:

> There must be in all this city area, provision for working class people, for people of low and middle income, to be able to reside in this area. No longer can unions only be content with battling around economic issues.... And so there will never ever … [applause] … be any reconstruction [applause].... There will never ever be any reconstruction, any rejuvenation, regeneration, of this area until such time as the residents receive iron clad guarantees that people in the low income brackets – workers – can afford to live in these areas. No longer are workers going to be driven out of cities.
>
> (Mundey 1973 in Fiske 1984, 48 min)

The commitment of the NSWBLF to advancing the rights of low-income residents in urban development processes is striking. These are not actions designed to improve the wages and conditions of members, but matters of urban policy and social justice. The ban on work in inner-city development was an exertion of union power, but the real issue as seen by the executive of the union was not the ideological battle between the builders, labourers and employers, but 'the rights of low income earners to live in the inner city at reasonable rentals' (Pringle in Fiske 1984, 45 min). While resident occupation was fragile (and illegal) throughout this period, the outcome of this support was that low-income workers continued to dwell in The Rocks while the urban renewal process had commenced. By supporting residents who, in the face of evictions, decided to stay in their homes, the green ban was therefore not only a call for resident involvement and situated urban renewal, it was the achievement of an alternative urban development process whereby urban renewal and the retention of affordable housing occurred simultaneously. In addition to preserving low-income housing, the green ban at The Rocks saw new innovations in participatory planning, set out next.

Experiments in participatory planning: 'The People's Plan'

The ways that iconic policies circulate between different cities has recently become a key area of interest in urban research. Recognising policy formations as relational achievements, this work reveals the diverse processes and practices through which policies travel across national borders (McCann and Ward 2011). The introduction of participatory planning approaches through the green ban at The Rocks illustrates the diverse ways in which policies can take hold. With the support of the union and in consultation with architect Neil Gruzman, The Rocks Resident Action Group indicated its intention to commission *The People's Plan* in October 1972. Developed by residents in association with other professionals, *The People's Plan* advocated the protection of the neighbourhood as a residential area with historical values, and for ongoing consultation between residents and the SCRA. *The People's Plan* was released in April 1973 and recommended that 'the operations of SCRA must be opened up, and structures set up for full and ongoing open consultation between the Authority and residents, tenants and interested citizens' (Roddewig 1978, 25).

While SCRA maintained a critical stance in public about the green ban and threatened low-income residents with legal action for remaining in their homes (Burgmann and Burgmann 1998) in its 1973 review, the Authority recommended an increase in the proportion of residential space and retention of historical areas. The following year, 107 families in social housing were permitted to stay in The Rocks with no increases in rents; surveys were undertaken by the planning and development authority to gain resident feedback; and a new advisory committee was appointed comprising residents, local businesses and representatives from SCRA. By 1975, SCRA and the NSW Housing Commission had reached an agreement to maintain low-income housing in the precinct through the proceeds of development elsewhere on the site. This funded construction of the nine-storey Sirius building, an icon of brutalist architecture, as an extension of the affordable housing stock in the precinct (Roddewig 1978, 26).

These changes were reflected across the city more widely, supported by a national policy shift with the Federal Whitlam government (ALP) protecting and expanding affordable housing in three of the green bans sites (Ruming *et al.* 2010). In response to pressure from residents maintained through the green bans, as well as the changed national policy framework, in 1974 the State government undertook a 'wide ranging investigation into land use planning' (Roddewig 1978, 108). The new Environmental Planning Bill (1976) was released soon after, focusing on resident input into planning processes, including: the requirement for councils to notify the public in advance of major development (Section 107); an extension of the right to object to all residents (not just adjacent land owners) (Section 110); and the guarantee of third-party appeal rights (Section 120) (Roddewig 1978). While rights were limited in the final legislation,[1] the State government nonetheless advanced wide-ranging planning and environment reforms including, by the end of the decade, the creation of the Land and Environment Court (NSW), such that any person could challenge planning decisions seen to breach the Environmental Planning and Assessment Act 1979 (Ryan 2002).

The green bans, then, not only advanced community-unionism, social justice and working-class and residents' rights, they also became a movement that influenced public policy in planning and housing in NSW. Reflecting on the green bans in 1998, Professor Pat Troy, Director for many years at the Urban Research Unit at the Australian National University, described the green bans in terms of their '(s)ubtle influence in transforming the culture of urban planning in ways that now evince greater sensitivity to environmental concerns, better appreciation of heritage, the need to publicise proposed developments well in advance and seek approval from the people affected' (cited in Burgmman and Burgmann 1998, 182–183). Combined with the protection of large pockets of low-income housing, greenspace and heritage buildings, the green ban at The Rocks effectively unbound housing/home from a regime of speculative office and commercial redevelopment and, through practices of striking, protest and occupation, moved social and ecological concerns centre-stage in urban renewal, with significant implications for housing and planning policies in NSW in subsequent years. Recognising the key roles of non-expert residents and workers in initiating these reforms, the next section explores the implications for housing and home.

Assembling an affordable housing policy

In *Living Room*, Jacobs and Smith (2008, 518) call time on the longstanding distinction between housing and home. To the extent that housing is placed on the side of 'markets, regulation and policy' and home on the side of 'meaning, memory and identity', this division is seen to conceal the processual, open-ended character of housing/home where emotional, material, regulatory and economic dimensions intersect. Today, there are (at least) two dominant conceptualisations that move beyond the binary of housing/home: the construction of housing markets and policies as products of emotional sensibilities and material practices (J.M. Jacobs 2006; Smith *et al.* 2006); and feminist and ecological constructions of the political and public nature of 'private' homes (Blunt and Dowling 2006; Crabtree 2006; Davison, this volume). In their feminist-inspired re-reading of the green bans, Anderson and Jacobs (1999) foreground the latter, emphasising the inherently public nature of the homes of social housing tenants, who were subject to the decision-making of state institutions; and the contrasting position of owner-occupiers at Kelly's Bush who sought to expand the borders of their private homes, to protect the public parkland in their neighbourhood. Their work thus shows the partial and contingent association of home with private space.

Given the influence of social movements on housing policy as set out above, I want to consider a third way in which the green bans troubles a clear distinction between housing/home. As we have seen, the green bans had significant implications for housing policy and practice. Yet housing policy formation, like many professional and government activities, is situated within traditions and histories of knowledge creation, problematisation and expertise (K. Jacobs 2006). In the context of marketised housing systems, Crabtree (2013) has shown that more-than-economic attachments to place, whether formalised through tenure diversity or sedimented through everyday practice, slip outside the speculative grammars of urban property development and housing regulation (see also Crabtree, this volume). Yet, in the case of the green bans, housing policy and implementation did indeed become the concern of those excluded from public decision-making. Combined with the force of the stilled building site, these actors who are routinely excluded from the development of housing policy and regulation moved centre-stage in the development of *de facto* policies that protected low-income dwellings (and more) in the context of urban renewal. Rather than an interest group influencing planning decisions, this is an organised coalition of residents and workers that manifested as an informal but powerful housing and planning power with material and discursive outcomes. The contribution that urban social movements make to thinking about housing/home, then, is to recognise that housing *policy* is a product of the decisions, actions and collaborations of non-expert residents and workers. Given the highly politicised nature of urban development in Sydney at the time, these collaborators emerge as an important resource for rethinking articulations of socially just housing in the context of marketised urban development.

It is precisely those elements that are 'outside' or (at least) only partially contained by a given system that, for Grosz (2001), comprise the locus of change and innovation. In the context of urban and housing policy in The Rocks, low-income residents, socially conscious workers and under-capitalised buildings were conceived as problems, 'impossibilities' and burdens. The extension and use of SCRA's powers to evict over 400 working-class tenants and demolish their homes contributed to the widening oppositions between working-class and elite neighbourhoods, as well as between the imagined future of housing (and planning) policies and the lived realities of citizens in everyday life. Residents and their homes were thus conceived as passive subjects, awaiting eviction and demolition so that the renewed Rocks could take its rightful place in 'Australia's greatest city'.

Despite this, all that was resistant within urban policy and development in Sydney at the time – the opinions and views of the construction workers, residents whose houses and neighbourhoods would be changed, buildings that were not turning profits and generating revenue – formed new connections and alliances in a reinvention of human capacity for dwelling beyond logics of property speculation and development. This is not merely residents and workers influencing the decisions of planning, but the wresting of the planning and urban development process away from the habits, modes of organisation and political compromises it was accustomed to. Yet the transformation is not an exclusively 'social' phenomenon. Rather it requires socio-material collaboration where residents, workers and buildings take on unexpected, wholly surprising functions and hybrid forms.

Recognising the key role of the stilled building site in both protecting housing and facilitating collaborative planning processes, the next section considers nonhuman agency in more detail.

Nonhuman collaboration: the role of the stilled building site

The protection of housing in The Rocks and the revisions to SCRA's redevelopment plan were both advanced through the withdrawal of labour and the stilling of the building site (including homes) by the NSWBLF. The effectiveness of this withdrawal is testimony to the socio-material alliances that comprise the urban development process. As noted elsewhere (see Cook 2016), resident calls for greater input into planning and housing policies were not only made through verbal and linguistic means – through, for example, speeches at rallies, depositions made to politicians or arguments and responses in industrial courts. Important as these expressions and their material impacts were, one of the distinguishing features of the green bans was that the call for a consideration of resident values and diverse attachments to place was also expressed through the materials and practices of urban development. Indeed, to the extent that the withdrawal of labour protected low-income housing and the rights of tenants to remain at home, the call is expressed through the building site itself: such that the green bans interrupted demolition, site clearance, preparation and construction. Beyond the placard, the deposition and the speech, the call is 'scattered' (Grosz 2001, 58) through the materials, places and practices of the urban development process. While the achievement of such a shift was the product of multiple actions including sustained resident protest, the democratisation of the union, the rise of New Left politics, and earlier industrial action, it also produced important transformations in both the call for greater input into planning and the function of the building site.

First, the boundary between *the call* for resident input into planning and housing policies and the *achievement* of resident input began to blur. To the extent that the stilling of the building site allowed residents to dwell in place and to negotiate with SCRA, the call was no longer simply a request, resident input had started to occur. Rather than a bounded, passive thing, the stilled building site – like the material world more broadly – was a resource for the formation of the new subject of the consulted resident. Withdrawing labour is not something unique to industrial disputes; it is how many contemporary contracts are negotiated between large commercial players. Research with resident objectors to higher-density housing also suggests that physical changes in development sites are more sensible to resident objectors than other types of written notification (see Cook *et al.* 2013). Yet the green bans moved beyond these more routine human–building interactions to tie the construction site to resident opposition.

Second, the building site was transformed. No longer a typical component in the accumulation of capital for large construction and development companies, the stilled building site becomes a transmission point for affordable housing and participatory planning in the centre of market-based urban development. This change does not emerge from a force that is external to housing in the context of market societies, but through the churning over of existing elements – including workers, residents and building sites – in new combinations (see Massumi 2002, 109). The green ban advances immeasurably through the withdrawal of labour as the building site is transformed into a system for the production of social and ecological values in urban development.

The green bans then were not only a cross-class coalition involved in civil action and the contestation of urban space. Through the internal churning of human and nonhuman elements, the movement provoked the evolution of the planning and housing system in NSW, wresting it from the logic of state-led speculation and opening it up to deliberative politics, diverse attachments to place and distributive justice. The outcome was not a distant utopia, a plan outside of a complex reality, but a wholly extraordinary event in the world as it is: a collaboration of workers, residents and buildings that ensured the carriage of social and ecological justice through the building sites of capitalist urban development.

Containing invention

> How it is possible to revel and delight in the indeterminacy of the future without raising the kind of panic and defensive counterreactions that surround the attempts of the old to contain the new, to predict, to anticipate, and incorporate the new within its already existing frameworks?
>
> (Grosz 1999, 16)

Grosz (1999) has argued that while events characterised by 'uncontainable change' and unforseeability can be understood as truly new, such indeterminacy is profoundly disruptive to existing economic, political and cultural orders. In contrast to predictable change, which is seen as a 'presumed social prerequisite',

> upheaval, the eruption of the event, the emergence of new alignments unpredicted within old networks threatens to reverse all gains, to position progress on the edge of an abyss, to place chaos at the heart of regulation and orderly development.
>
> (Grosz 1999, 16)

Such events become the object, not only of suppression and regulation, but of attempts to be drawn into the 'fabric of the known' (Grosz 1999, 16). Indeed, it would be misleading to assume that the event of the green bans lead to an ideal city. By 1973 Mundey reported developers had offered him 10 per cent of the profits on sites in exchange for removing the green bans as well as penthouse apartments in new high-rise (Mundey 1981). At the same time, following concerns raised by police he reported that a contract for his assassination had been drawn up. From a legal perspective, occupying the building site was already subject to new rounds of enclosure through revisions to the Summary Offences Act. By 1974 attempts to deregister the NSWBLF were initiated by the Master Builders' Association, completed two years later by the Federal Builders Labourers Federation; a decision that effectively ended green bans on all building sites. Darcy and Rogers (2015, 5) argue that the potential for community-based unionism today is limited through the tightly controlled business of workplace bargaining. This includes Federal controls on workplace organisation and heavy penalties for 'secondary boycotts'. This is not only a penalty for engaging in disputes outside of wages and conditions, it is also a financial disincentive to foster community-based unionism.

However, throughout the 1970s, participatory planning mechanisms and protection of heritage and affordable housing were imported directly into the planning process at The Rocks and by the end of the decade into NSW planning law more broadly (Cook 2007). Consider how, through the course of the green bans, the SCRA began to consult with residents, seeking out resident opinion and responding to concerns by preserving and increasing the stock of affordable housing in The Rocks; how, by 1979, resident and council objection was solicited through formal planning processes with the possibility of a court appeal for unlawful decisions; and that heritage and environmental controls were developed alongside new planning processes. All of these strategies work to manage the encounter between social movements and planning and housing policy, to standardise, limit and control the reach and scope of subsequent encounters.

Conclusion: assembling affordability

The green bans were an extraordinary event – not for their consistency with existing ideals of working-class action, union protest or urban social justice – but for their reworking of established categories, groupings and forms. Whether labourer, communist, social housing resident, or owner-occupier, established identities were left behind in the virtual city where workers and residents became planners and housing authorities, and where the planning system was, for a time at least, immersed in the rich variability of urban life. The event fundamentally changed the planning process in NSW so that, in order to contain opposition to neighbourhood change, the participatory planning approaches already in use among the green ban activists became the subject of new planning practices, policies and laws.

The green bans also challenge the binary of housing and home in that the world of policy was open to residents, citizen-led experimentation and diverse attachments to place. By drawing on philosophies of materialism, and virtuality, the green bans are positioned as a social movement that manifested as an informal but powerful institution of urban policy, housing and planning which unsettled, (partly) reordered and established new policy frameworks for urban planning and housing in subsequent decades. While the transformation of materials, society and space through the period of green bans was a contingent and fragile achievement, the event nonetheless demonstrates the eminent capacity for innovation within housing systems, fashioned not in a distant future, but within the constraints of urban dwelling in the present.

Note

1 NSW has never had, for instance, as extensive third-party objection and appeal rights as Victoria (see Cook *et al.* 2012).

References

Amin, A. 2012. *Land of Strangers*. Polity Press, Cambridge.
Anderson, K and Jacobs, J. 1999. Geographies of publicity and privacy: residential activism in Sydney in the 1970s. *Environment and Planning A* 31: 1017–1030.
Birmingham, J. 2000. *Leviathan: The Unauthorized Biography of Sydney*. Vintage press, Milsons Point.
Blunt, A. and Dowling, R. 2006. *Home*. Routledge, London.
Burgmann, M. and Burgmann, V. 1998. *Green Bans Red Union*. UNSW Press, Sydney.
Cook, N. 2007. Conflict and consensus in the redevelopment of property. Unpublished PhD Thesis, Macquarie University.
Cook, N. 2016. The virtual city: the green bans, participatory planning and the space of the new. *Environment and Planning D: Society and Space* (in review)
Cook, N. Taylor, E., Hurley, J. and Colic-Peisker, V. 2012. *Resident Third Party Objection and Appeals Against Planning Applications: Implications for Higher-Density and Social Housing*, Final Report No. 197. Australian Housing Urban and Research Institute, Melbourne.
Cook, N. Taylor E. and Hurley, J. 2013. At home with strategic planning: reconciling resident views with everyday processes of territorialisation. *Australian Planner* 50(2): 130–137.
Crabtree, L. 2006. Disintegrated houses: exploring ecofeminist housing and urban design options. *Antipode* 38(4): 711–734.
Crabtree, L. 2013. Decolonising property: exploring ethics, land, and time, through housing interventions in contemporary Australia. *Environment and Planning D: Society and Space* 31: 99–115.
Daly, M. 1982. *Sydney Boom, Sydney Bust: The City and Its Property Market, 1850–1981* Allen and Unwin: Sydney.
Darcy, M. and Rogers, D. 2015. Place, political culture and post-Green Ban resistance: public housing in Millers Point, Sydney. *City*. DOI: 10.1016/j.cities.2015.09.008.
Fiske, P. (producer and director) 1986. *Rocking the Foundation* [Documentary Film]. Australian Film Finance Corporation, Australia.
Freestone, R. 2000. Planning Sydney: historical trajectories and contemporary debates. In J. Connell (ed.) *Sydney: The Emergence of a World City*. Oxford University Press, Melbourne.

Grosz, E. 1999. *Becomings: Explorations in Time, Memory, and Futures.* Cornell University Press, New York.

Grosz, E. 2001. *Architecture from the Outside: Essays on Virtual and Real Space.* MIT Press, Boston, MA.

Hardman, M. and Manning, P. 1975. *Green Bans: The Story of an Australian Phenomenon* Australian Conservation Foundation, East Melbourne.

Hasham, N. and Johnstone, T. 2014. Sssh, it's top secret: Millers Point house goes up for auction. *The Sydney Morning Herald,* 21 August 2014.

Healey, P. 1997. *Collaborative Planning for Fragmented Communities.* Macmillan Press, London.

Iveson, K. 2014. Building a city for 'the people': the politics of alliance-building in the Sydney Green Ban Movement. *Antipode* 46(4): 992–1013.

Jacobs, J.M. 2006. A geography of big things. *Cultural Geographies* 13: 1–27.

Jacobs, J.M and Smith, S.J. 2008. Living room: rematerialising home. *Environment and Planning A* 40: 515–519.

Jacobs, K. 2006. Discourse analysis and its utility for urban policy research. *Urban Policy and Research* 24(1): 39–52.

Jacubowicz, A. 1984. The Green Bans movement: urban struggle and class politics. In J. Halligan and C. Paris (eds), *Australian Urban Politics: Critical Perspectives.* Longman Cheshire, Melbourne.

McCann, E.J. and Ward, K. 2011. *Mobile Urbanism: Cities and Policymaking in the Global Age.* Minneapolis, MN: University Minnesota Press.

Massumi, B. 2002. *Parables for the Virtual: Movement, Affect, Sensation.* Duke University Press, Durham, NC.

Mundey, J. 1981. *Green Bans and Beyond.* Angus and Robertson, Sydney.

Roddewig, R.J. 1978. *Green Bans: The Birth of Australian Environmental Politics.* Hale and Iremonger, Sydney.

Ruming, K., Tice, A. and Freestone, R. 2010. Commonwealth urban policy in Australia: the case of inner urban regeneration in Sydney, 1973–75. *Australian Geographer* 41(4): 447–467

Ryan, P. 2002. Court of hope and false expectations: land and environment court 21 years on. *Journal of Environmental Law* 14(3): 301–315.

Sandercock, L. 1977. *Cities for Sale.* Melbourne University Press, Melbourne.

Shaw, K. 2008. Gentrification: what it is, why it is, and what can be done about it. *Geography Compass* 2(5): 1697–1728.

Smith, S.J., Munro, M. and Christie, H. 2006. Performing (housing) markets. *Urban Studies* 43(2): 83–98.

Thomas, P. 1973. *Taming the Concrete Jungle: The Builders Labourers Story.* Australian Building Construction Employers and Builders Labourers Federation, Sydney.

12 Interstitial housing space
No centre just borders

Wendy Steele and Cathy Keys

Introduction

The informal and taken-for-granted spaces in the margins of housing present a vital site for the creation, maintenance and preservation of the house/home. This chapter reconceptualises housing/home in relation to interstitial space, highlighting the ways that such spaces perform multiple roles in sustaining the continuity of dwelling against everyday social and ecological uncertainties. Our intersection with the book's overarching theme of housing and home unbound is an exploration of interstitial space as a site of indeterminacy through which the boundaries of housing/home are unsettled.

Interstitial housing space focuses critical attention on those places and practices whose basic nature falls within and between the familiar boundaries of dwelling and home. Interstitiality implies a reconceptualisation of space that lies somewhere between the 'here' and 'there', 'us' and 'them', and between notions of fixity and certainty (Gupta and Ferguson 1992). Previous studies on the materiality of home focus on housing (as material, fixed space) and home (as felt or emotional space). However, a separate literature on informality also emphasises the processes of dwelling and homemaking as well as the capacity of materials and spaces to take on unexpected roles. This includes bringing to the surface different representations of the domestic, the meaning of home and the uses of housing types as dwellings for different cultural groups such as Indigenous and settler Australians.

Drawing on two informal housing spaces in Australia – 'under the house' in the Queensland colonial vernacular, and the yulka found in remote Aboriginal camps and housing in Central Australia – this chapter makes a claim for the interstitial. Such spaces, precisely because of their temporality and flexibility, are shown to be vital in sustaining home (as a space of comfort and security) and living environments against social and ecological uncertainties. In doing so, the chapter advances a conceptualisation of home that recognises the informal, flexible materials and spaces through which housing/home is sustained. It also highlights the importance of indeterminacy and informality in housing research in market and colonial settler societies.

Interstitial space: materiality + informality

There is a considerable literature within housing studies on the meaning of home and material practices (Blunt and Dowling 2006; Cruz 2008; Easthope 2004; Miller 2001; Pink 2006). A number of housing researchers have focused on material practices as a means by which to 'unsettle ideas about home and domesticity by questioning what might at first appear to be familiar and mundane' (Blunt 2005, 505). The subjective nature of housing as dwelling, for example, is understood to be as much about ordinary ways of being and finding meaning in the world as a physical place or material economic asset. In this sense, dwelling is about settlement; 'about moving in the environment, about making and keeping community, about finding our place' (King 2008, ix).

Studies on the materiality of home have already done much to blur the boundaries of housing and home. Research by Daniel Miller (2001), unsettles the border between the material/immaterial through a focus on material objects in the home. He argues that the material practices of home open important yet neglected windows for fine-grain social and cultural analysis of diverse domestic housing practices. His work shows that the achievement of feeling at home hinges on a range of non-human collaborators. These objects – or materials – are vital in the achievement of home. They co-produce the feeling of security and comfort that we often associate with home. Conceptually and methodologically, Miller offers a strong narrative emphasis on the importance of the 'generic ordinariness of objects' as a means of ultimately rescuing 'the humanity' of ordinary people.

Nicole Cook *et al.* (2013) argue that it is important to acknowledge that objects, materials and things are able to play a constitutive role in the social life of housing and home. For Greg Noble (2004), people construct and stabilise their identities and families in and through a set of home materials around which practices of remembering can help solidify otherwise distant personal connections – photographs, mementos and special objects are important here. Noble argues that as we accumulate objects, we accumulate *being* based on research into material culture in the home. He explores the complexity and continuity of everyday life, grounded in the material objects and spaces of the home. Divya Tolia-Kelly (2010) points to the different ways that home materials can help refract distant homelands into the present – in a material sense to unsettle the forces of colonial hostility for British Indians in London. Using both visual and material culture, she explores interstitial spaces of memory and belonging presented as mobile and as politically charged, with as much meaning as the more formal spaces of race in a postcolonial landscape.

Rematerialising housing by moving beyond the home/housing binary as outlined by Jacobs and Smith (2008) intersects with key notions around housing and informality. The potential of informal practices is outlined by Cruz (2008), whose work critically explores 'the ways in which informal settlements creatively reuse "waste" material and make flexible spaces with overlapping programs' through everyday informal housing practices (Cruz 2014a, 1). Within the broader geopolitical border context his focus is the micro-scale of home and neighbourhood and the multifarious ways in which this bottom-up civic approach to housing and development supports 'informality and alternative social organizational practices' (Cruz 2014b, 1). Bottom-up solutions and hybrid spatial practices he argues produce new interpretations of social and infrastructure, property and citizenship through the interplay between interstitial space and everyday informal housing practices.

The literature on interstitial space emphasises the ways that spaces and materials get taken up in a wide range of everyday practices that often unsettle the intended uses of such spaces/materials. These spaces and materials have a type of indeterminacy – they can go this way or that, depending on the context. As Colin McFarlane (2011, 659) describes, this involves 'the tinkering and tweaking of urban assemblages over time'. For Sankalia (2008, 29), interstitial spaces such as these are found 'hidden between the facades of buildings – the "slots", nooks, and crannies of the City'. Here, the interstitial reflects 'more than a mere site – but a small, hidden unit in the urban fabric' (Rahmann and Jonas 2011, 5) and in doing so offers alternatives and a point of differential from the predictable spaces of consumption 'especially in fluctuating, dynamic, ever-changing network' urban environments (p. 9).

Interstitial housing space exists therefore on the margins or edge worlds of domestic life, and offers a contribution to the rich cross-disciplinary theorising in the social sciences around the importance of interstitiality as alternative spaces and places in contemporary capitalist times (Cruz 2008; Rahmann and Jonas 2011; Zizek 2006). Working in the cracks of the established housing order, the interstitial offers a focus on those spaces and places that avoid 'rigid codification' (Barron 2014). A form of in-between or loose housing space that is both complex and contradictory yet filled with the creative possibility that comes from being located at the edges and peripheries of domicile (Steele and Keys 2015). This is the 'liveliness of space' referred to by McFarlane (2011), whereby informal housing practices are indeterminate but not always simply ad hoc, enriched by possibilities and iterative processes of hidden learning and adaptation that brings 'a wider spatial matrix to how dwelling might be understood, important considering the bounded, leaden-footed connotations that dwelling can sometimes buy into' (p. 662).

Housing borderlands: the Australian urban context

Interstitial space is a site of informal or taken-for-granted housing possibility that blurs the boundaries of housing (as material) and home (as lived). To illustrate this nexus we draw on two different examples from within the Australian housing context: 'close to the wind-break' found in living environments including housing and camps created by Warlpiri-speaking Aboriginal people living in remote desert communities in the centre of Australia; and 'under the house' of raised timber-framed housing built throughout the Australian state of Queensland and in it's capital city of Brisbane.

Both the case studies are built up primarily from secondary sources and draw on bricolage social science methodology that is part auto ethnography, part literature review and part critical narrative (see, for example, Law 2004; Steele 2011). The Central Australian case study was extended by secondary sources within the field of social anthropology, where an extensive body of scholarship has centred on both the pre-contact and contemporary life of Warlpiri people of the Tanami Desert of Central Australia since the early 1960s. Anthropologist Yasmine Musharbash's doctoral thesis (2003) and later book *Yuendumu Everyday: Mobility: Intimacy and Immediacy in an Australian Aboriginal Settlement* (2008) are key sources as they demonstrate both continuity and change in the use of Warlpiri domiciliary behaviour beyond the findings of doctoral research into Warlpiri women's housing need performed by Keys in the 1990s (see Keys 1999). Musharbash explored relationships between meaning and domestic space, finding that Warlpiri camps, like the house were richly embedded with social meaning and reminded us of the need to more critically examine our understanding and use of terms such as housing/home in culturally diverse settings (Musharbash 2008, 26–45). Material concerned with 'under the house' was drawn from the field of architectural history focused on the building typology and regionally specific features of the Queensland house (Bell 1984; Evans and The National Trust of Queensland 2001; Riddel 1990; Roderick 2004).

These Australian domestic housing examples have been chosen because they illustrate in quite different ways the everyday and material practices that exist in these spaces, including: the possibilities inherent with understanding informal housing practices located 'under the radar' for subverting housing norms; the ways these spaces and practices shift and change at the domestic private scale over the housing life-course; and the importance of interstitial space in quite different cultural contexts. Together the cases work to advance understandings of materiality and housing/home and challenge the disciplining of housing space in market-based settler-societies – towards intensive development and marketisation.

Subverting housing norms: the slots, nooks and crannies

Interstitial housing space at the domestic scale often refers to both physical and conceptual spaces 'at the edge and peripheries of buildings like pavilion, patio, verandas, corridors, balconies, doors and windows' (Chan 1997, 2). For Heathcote (2012), the meaning of home has become so familiar that we have forgotten how to appreciate the detail of domestic features and practices – the elements of the everyday that through careful exploration and interrogation offer deep wells of meaning and insight into how we live in private domestic spaces.

A form of interstitial domestic living space occurs in Central Australia, where the Warlpiri-speaking Aboriginal people live in towns and communities located in desert communities in the Northern Territory. The yunta (or wind-break) was a living space used in favourable weather conditions predominately between sunset and sunrise behind a wall where sleeping, cooking and eating and storage functions occurred (Keys 1997). In an outdoor context, yunta was physically made up of a barrier wall (which could be short or tall), usually running in a north–south direction, sometimes with end walls creating a protected sleeping area to the west of this barrier and fires for warming or cooking. Yunta were found in camps but also in bedrooms, on verandas, or in house yards (Keys 1997, 1999; Musharbash 2008) (see Figure 12.1).

Where Warlpiri people lay their heads in the yunta had a practical role. It's where people stored things. Within the yunta, personal belongings continue to be stored at the head of an individual's sleeping area between the windbreak and the sleeper's head (Keys 1999; Musharbash 2003). Musharbash (2003, 49) describes the nature of these things at the turn of the twenty-first century as 'essential items, such as water, bottles, matches, wallets, keys, handbags, and whatever else is important to the sleeper'. This

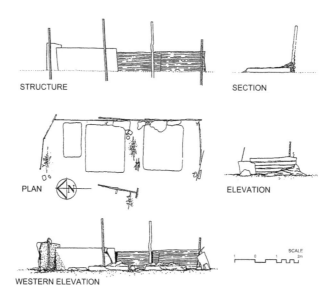

Figure 12.1 Plan, section and elevation of yunta (wind-break), Nyrripi, May 1994.
Source: Keys 1999, 177.

interstitial space was known as *yalka*, which the Warlpiri dictionary translates as 'close to the wind-break' (Musharbash 2003). If there was no windbreak, yalka was the space just above or under a pillow.

By contrast, domestic storage 'under the house' is an interstitial space common to the elevated 'Queensland house', the most widely constructed house type built in the state of Queensland until the Second World War (Watson 1981, 132). Early Queensland houses in low-lying areas were elevated on stumps, allowing floodwaters to pass under the house and 'not sweep it away' (Newell 1974, 39) The practice of raising these timber-framed houses up on a grid-work of hardwood timber stumps protected the softwood elements of houses from pests like termites; improved ventilation into living spaces (Bell 1984; Roderick 2004); mitigated flooding of living areas; and provided opportunities to use below-floor spaces created between the stumps for storage and service work spaces (Riddel 1994).

Storage of tools and associated workshop activities were an important use of the space 'left' under Queensland housing (Evans and The National Trust of Queensland 2001, 33). Laundry tasks were often done under the house and lines strung out between the stumps to dry washing during annual periods of wet weather. Personal possessions like gardening tools, sports equipment, children's toys and old furniture might be stored here (see Figure 12.2).

A laundry bench, washtubs, buckets, water taps and devices for washing and wringing often defined a 'women's work' area of 'under the house'. This wet space was positioned next to rainwater collection tanks, with regular periods of drought requiring close management of water usage. The garden benefited from recycled washing water. Aside from providing a shady retreat space protected from rain, harsh sunlight and in summer oppressive heat, under the house and between the stumps was a popular play area for children where they could play out of sight but within 'earshot'. This play area has a very strong spatial experience, that of 'being pressed between the house and the natural ground' (Wall 2004, 23). It was an interstitial space where children established lived understandings of place. The shadowy qualities of this space gave occupants the dual abilities to survey the outside world but also refuge and the ability to hide in the shadow of the floor above and the fringe of timber skirting its edges (Wall 2004).

Memories lived in this space (Bachelard 1994; Miller 2001; Wall 2004) and material objects associated with memories were stored under the house. No longer required 'upstairs', photographs, unseasonal clothing, old books and furniture were packed away under the house. Interstitial housing space lies outside of formalisation, a space of experimentation enabled by capacities of materials to enter into multiple relations and to be taken up in a wide range of practices and contexts. It is the indeterminacy of these materials and the diverse practices they enable that contributes most to the achievement of home.

Figure 12.2 'Under the house', Redhill, Brisbane.
Source: Keys 2013.

Shifting housing margins over the life course

Building on current understandings of materiality and house/home, the importance of the indeterminacy of materials in shoring up housing/home against the odds calls for a greater consideration and engagement with these in-between spaces. In *Home Truths*, Sarah Pink (2006) advocates for the need for a better understanding of how housing practices are constituted and shift over the housing life-course to emphasise the need to engage more deeply with the fine grain of everyday lived and material practices (e.g. walking, possessions). Pink (2006) argues that it is in the mundane, the hidden and the ordinary that processes of renewal, difference and change can be found.

While domestic mobility is still a strong feature of Warlpiri contemporary life, house use has mostly replaced the use of traditional vernacular architecture and the occupation of self-constructed camps (Musharbash 2008, 10). Over time, concentrated settlement has decreased access to the diversity of traditional storage areas such as on the roof of shade structures or in trees, and reduced the number of 'hiding places' and opportunities to store personal things for longer periods of time. Warlpiri people living in Yuendumu have been presented with a range of house types based on mainstream Western house forms and living practices found in urban environments by Australian governments since the 1950s (Keys 2000). By the 1990s, Warlpiri women living in jilimi (women's camps) with access to these types of housing preferred to sleep outside in good weather in a yunta or on verandas up against the sides of houses, using internal rooms for storage.

In poor weather, internal rooms in houses were used for sleeping. Musharbash (2008) similarly describes the social and cultural properties of room use a decade later, noting a preference for externally oriented behaviour (which she calls yard-orientedness) in good weather and utilisation of rooms in houses primarily for storage. Regardless of the location of women's sleeping places, they continued to store important personal possession in the interstitial space between their head and a wall or the yalka.

In the 1990s, long-term storage of possessions was a requirement of Warlpiri living environments that was not being met (Keys 1999). Women with access to housing stored clothing and important personal items in cardboard boxes, plastic bags, carry bags and suitcases. There were only two houses with cupboards occupied by Warlpiri women living in Yuendumu at this time and the houses were highly regarded, primarily for this reason alone. Open shelves in kitchen spaces of some houses were not used because they could not be supervised and people could not lock up their possessions.

Ongoing utilisation of yalka observed by Musharbash (2008, 43) illustrates some of the spatially and culturally situated practices around storage use that continued, despite an increase in the diversity and volume of material possessions and changing housing opportunities.

In the last 20–30 years growing urban densities and volumes of personal material possession have placed increasing pressure on the spaces left 'under the house' of the Queensland house. The result has been more physical occupation and building-in of this space, replacing the flexible and semi-permanent nature of activities with defined programmed space and permanent occupation. As older houses have been raised, the ground plain excavated and the sub-floor area built-in, the interstitial space of 'under the house' in Brisbane is gradually disappearing from the urban fabric. This has occurred alongside the reduction of yard spaces and the loss of garden sheds (another traditional storage place) from inner city suburbs (Hall 2010).

As inner city blocks of land have been subdivided, self-contained units, laundries, home offices, children's play rooms, poolrooms, art studios and the multimedia rooms have been built-in under existing residential floor plans. Built-in cupboards have absorbed some of the 'treasure' previously found 'under the house'. While there has been an increase in the diversity of spaces and activities possible by building-in, the nature of these spaces has changed and became more internally oriented. Floor surfaces are no longer dirt but timber, concrete or tiles. The edge walls are no longer semi-transparent and permeable but solid and fixed. Air flow and temperature are often mechanically controlled and filtered through reverse cycle air conditioners and daylight is supplemented (or replaced) by electric lighting.

Even though people were more grounded (literally occupying the ground plain), renovating and occupying 'under the house' has resulted in an increase in filters and layers between people and their environment that had not been there when this space had an open or semi-filtered relationship to the elements. During the 1980s and 1990s Brisbane experienced prolonged periods of drought and residents developed a false sense of security with regard to the cities location on a floodplain. As families grew and became more affluent, houses were lifted and 'left over' or 'empty' spaces 'under the house' were fitted out and renovated to absorb the changing needs of contemporary life.

Interstitial housing space and cultural difference

Following Brown (2008, 15), interstitial housing space can be understood as the 'permeable, multilayered ambiguous filtered edge ... about change in social and cultural values of a society – changes in the way we relate to each other and way we relate to environment'. For King and Dovey (2013), it is through interstitial spaces and everyday practices that transitions to different social/cultural conditions are able to shift and rise. Yiftachel (2012) links the idea of interstitial space such as informal settlements with other struggles from below that seek to create particular cultural and ethnic spaces of identity. Interstitial space (grey space) involves 'a slow encroachment of the ordinary', working in *the cracks* of the established order (Yiftachel 2012, 152).

Historically, all levels of Australian government have been challenged by housing provision for Indigenous Australians and have been open to criticism of applying 'an ethnocentrically European model of housing need rather than a client's perceptions of their needs' (Thompson 2004, 273).

Exploring cultural differences in an understanding of terms like home and camp, there is a significant difference between the non-Aboriginal term 'wind-break' that focuses on a single physical characteristic, a wall, and the Warlpiri term 'yunta', which is used to describe a living environment with social and physical properties (Musharbash 2003). The yalka can be understood as an important interstitial space located within this living environment, providing a culturally appropriate place in which to temporarily hide and store important personal possessions.

Within Warlpiri communities in Central Australia an important function of housing is storage. However, the culturally specific importance of, and practices associated with, storage tends not to be successfully addressed in contemporary house designs and has often resulted in a 'lack of fit' between what people need and the housing made available to them. Environmental psychologist Joseph Reser (1979, 89), who considered approaches to Aboriginal housing in the 1970s, noted:

> The inability to store things of value in a person's own home is a fundamental and radical departure from the traditional situation in which an 'unlawful' borrowing of a possession or even entry into another man's dwelling might result in being speared. One of the principle functions of a traditional shelter was storage.

In Central Australia it can be argued that one of the reasons why government-provided housing has failed to meet Aboriginal residents' needs to store possessions is linked to the inability of occupants to hide or securely keep possessions. Before contact with European settlers, Warlpiri people did not have more possessions than they could carry and store in their yalka. In the absence of a pillow, sleepers rested their heads on their personal possessions in the yalka (Musharbash 2003). Hiding possessions or removing them from easy surveillance decreased the demand placed on them (see Smith 1996).

While ongoing maintenance of demand-sharing behaviour places limitations on the amount of possessions that can be accumulated (Musharbash 2003; Peterson 1993, 2013), the need for storage space has increased and will continue to increase over time. An understanding of sharing behaviour and the social, cultural and spatial properties of storage of personal possessions in the interstitial space of the yalka (close to the wind-break) is valuable. It provides strong clues to addressing some of the storage requirements of residents with a respect for cultural difference. Built-in cupboards and shelving in mainstream housing may well meet the storage needs of many Warlpiri people living in contemporary settlements, but designing spaces that support the continuing cultural practice of storing important personal possessions in yalka might also include consideration of: the edges against which people might orientate themselves to sleep; the height at which they might sleep (elevated on a bed frame or on the ground plane); opportunities to make temporal storage spaces under and beside the sleeper's head at varying heights from the ground and a concern for opportunities to discreetly hide/screen possessions and then visually survey access to such places during the day; and the provision of storage spaces of a range of sizes that are secure and lockable.

Misunderstanding the complex balance of functional, social, cultural and spatial properties and importance of interstitial space can also be illustrated with Queensland house experience of 'under the house'. Between November 2010 and January 2011 Brisbane experienced a natural disaster through a period of widespread flooding. Slow-moving floodwater inundated many inner-city houses located in low-lying areas along the Brisbane River. Building fabric in the flood path was saturated and damaged alongside people's belongings. Personal possessions were destroyed and dumped into huge landfill operations in the 'clean-up' carried out over the following weeks and months. This climatic event resulted in widespread and significant personal, financial and emotional loss and waste production, with residents forced to rethink their relationship with the elements in terms of their housing.

A direct local government planning response to the Brisbane floods has been to redefine the flood level and to allow increased residential building heights in flood-prone areas, permitting the lifting of houses up to another storey higher (Skinner 2011). At the time of writing it was possible to walk along creeks in inner-city suburbs and low-lying older suburbs and see older houses or new two-storey residentials being raised up above the new flood level on steel stumps. The ground plane is being reinstated with visible 'under the house' from one side to the other, revealing storage of building materials used in the renovations. These developments mark a return of the interstitial space 'under the house'. Users are familiar with these types of 'uninhabitable' spaces, and it is just a matter of time before the next generation of children crawl in 'under the house' and lay down their own set of spatial and cultural storage, practices and memories.

Conclusion

Interstitial space has yet to attract considered interest in housing research in Australia. Moments and spaces that do not yet add up to a fully formed owner-occupied home are often interpreted as an interruption on the way to home ownership, and the creation of the homeowners as an investor subject. Whether and how such spaces open out other possibilities, particularly in socially and ecological uncertain contexts, and with an eye to notions of cultural difference, has yet to be explored. This literature on informality resonates with the increasing interest in interstitial space more broadly. This chapter highlights the importance of better understanding conditions of indeterminacy that comes with interstitial space in current understandings of materiality and house/home, even camp, and a greater level of engagement with the 'liveliness' of these in-between informal spaces.

We argue that the interstitial lens invokes the transformative potential of the margin as a critical space of housing, more often than not rendered invisible due to their informal form and private nature. It is therefore significant for the opportunities afforded to: identify alternatives, differences and opportunities to subvert mainstream housing space/form foci by investigating practices operating in the margins and the ways in which they shift over the housing life-course; focus on uncovering micro housing patterns and everyday politics and practices that can form the basis of counter projects and strategies to the regulated norm; and shed critical light on housing and space, including, but not limited to, aspects of rhythm, shadow and mystery.

Interstitial space is not a 'neutral grid', but one in which cultural difference, historical meaning and social practices are inscribed and actively made through everyday human practices (Gupta and Ferguson 1992). As illustrated through the two Australian cases within the private sphere of dwelling and home, different spatial practices emerge in the cracks of interstitial space in response to broader social and environmental pressures. Where the properties of this interstitial space are valued and understood by users in the case of the 'yalka' beside the windbreak, we see continuity despite considerable cultural change and even despite the 'improvements' of designers and housing providers. In contrast we see a devaluing of the flexible uses of interstitial space 'under the house' when these historical uses have been made to compete with the demands of greater urban densities and more programmed and fixed occupation.

Greater recognition of people's use of interstitial spaces in domestic settings (regardless of social or economic status) is critical to contemporary housing provision that meets people's existing and future dwelling needs. Yet housing research is largely silent in terms of: understanding interstitial housing space as a relational construct around public/private, mobile/static or renting/owning, etc.; the cultural context, nature and role of interstitial spaces both inside and outside the home, such as where children play, where residents chose to store possessions/memories, rest, socialise or seek solitude; the interstitial spaces between houses which tend to be causes of conflict and confrontation, such as between landlords and tenants, between neighbours and family members; and the transformative potential of interstitial housing as everyday processes of creative social and cultural practice. It is within this context that interstitial housing practices take on renewed logic and meaning. No centre, just borders amidst the flux, flow and indeterminacy of housing re-materialisation and re-bordering.

References

Bachelard, G. 1994. *The Poetics of Space*. Beacon Press, Boston, MA.

Barron, P. 2014. At the edge of the pale. In M. Mariani and P. Barron (eds) *Terrain Vague: Interstices at the Edge of the Pale*. Routledge, New York.

Bell, P. 1984. Miasma, termites and a nice view of the dam. In B.J. Dalton (ed.) *Lectures on North Queensland History*. James Cook University of North Queensland, Townsville.

Blunt, A 2005. Cultural geographies: cultural geographies of home. *Progress in Human Geography* 29: 505–515.

Blunt, A. and Dowling, R. 2006. *Home*. Routledge, Abingdon.

Brown, R. 2008. *Redefining the Edge: Reinventing Traditions for the Modern World*. Traditional Dwellings and Settlements: IASTE Working paper series. Centre for Environmental Design Research, Berkeley.

Chan, F.K.K. (1997). Potential of interstitial spaces in singapore's high rise public housing. Bachelor of Architecture Thesis, University of Queensland, St. Lucia.

Cook, N., Smith, S.J. and Searle, B.A. 2013. Debted objects: homemaking in an era of mortgage-enabled consumption. *Housing Theory and Society* 30(3): 293–311.

Cruz, T. 2008. Border tours: strategies of surveillance, tactics of encroachment. In M. Sorkin (ed.) *Indefensible Space: The Architecture of the National Insecurity State*. Routledge, New York.

Cruz, T. 2014a. *Spatial Agency*. Retrieved on 19 March 2014 from www.spatialagency.net/database/why/political/estudio.teddy.cruz

Cruz, T. 2014b. *From the Global Border to the Border Neighbourhood*. Retrieved on 19 March 2014 from www.california-architects.com/en/estudio.

Easthope, H. 2004. A place called home. *Housing, Theory and Society* 21(3): 128–138.

Evans, I. and The National Trust of Queensland. 2001. *The Queensland House: History and Conservation*. Flannel Flower Press, Mullumbimby.

Gupta, A. and Ferguson, J. 1992. Beyond 'culture': space, identity, and the politics of difference. *Cultural Anthropology* 7(1): 6–23.

Hall, T. 2010. *The Life and Death of the Australian Backyard*. CSIRO Publishing, Canberra.

Heathcote, E. 2012. *The Meaning of Home*. Frances Lincoln, London.

Jacobs, J. and Smith, S. 2008. Living room: rematerializing home. *Environment and Planning A* 40: 515–519.

Keys, C. 1997. Unearthing ethno-architectural types: categories of societies, meanings and properties of yunta (wind-breaks) of Warlpiri single women's camps. *Transitions*, February/March: 20–29.

Keys, C. 1999. The architectural implications of Warlpiri jilimi. Unpublished Doctorate of Philosophy, University of Queensland.

Keys, C. 2000. The house & the yupukarra (married people's camp): a history of housing in Yuendumu 1950s–1990s. In P. Read (ed.) *Settlement: A History of Australian Indigenous Housing*. Aboriginal Studies Press, Canberra.

King, P. 2008. *In Dwelling: Implacability, Exclusion and Acceptance*. Ashgate, Aldershot.

King, R. and Dovey, K. 2013. Interstitial metamorphosis: informal urbanism and the tourist gaze. *Environment and Planning D* 3: 1022–1040.

Law, J. 2004. *After Method: Mess in Social Science*. Routledge, Abingdon.

McFarlane, C. 2011. The city as assemblage: dwelling and urban space. *Planning and Environment D* 29: 649–671.

Miller, D. (ed.). 2001. *Home Possessions: Material Culture Behind Closed Doors*. Berg, Oxford.

Musharbash, Y. 2003. Warlpiri sociality: an ethnography of the spatial and temporal dimensions of everyday life in a Central Australian Aboriginal settlement. Unpublished Doctorate of Philosophy, Australian National University.

Musharbash, Y. 2008. *Yuendumu Everyday: Contemporary Life in Remote Aboriginal Australia*. Aboriginal Studies Press, Canberra.

Newell, P. 1974. The origins and development of the single family house in Queensland. Second draft for submission to Board of Architects of Queensland Award.

Noble, G. 2004. Accumulating being. *International Journal of Cultural Studies* 7: 233–256.

Peterson, N. 1993. Demand sharing: reciprocity and the pressure for generosity among foragers. *American Anthropologist* 95(4): 860–874.

Peterson, N. 2013. On the persistence of sharing: personhood, asymmetrical reciprocity, and demand sharing in the Indigenous Australian domestic moral economy. *The Australian Journal of Anthropology* 24(2): 166–176.

Pink, S. 2006. *Home Truths: Gender, Domestic Objects and Everyday Life.* Oxford: Berg.

Rahmann, H. and Jonas, M. 2011. Void potential: spatial dynamics and cultural manifestations of residual spaces. In M. Mariani and P. Barron (eds) *Terrain Vague: Interstices at the Edge of the Pale*, Routledge, Abingdon.

Reser, J. 1979. A matter of control: Aboriginal housing circumstance in remote communities and settlements. In M. Heppell (ed.) *A Black Reality: Aboriginal Camps and Housing in Remote Australia.* Aboriginal Studies Press, Canberra.

Riddel, R. (1990). Sheeted in iron: Queensland. In T. Howells and M. Nicholson (eds) *Towards the Dawn: Federation Architecture in Australia 1890–1915.* Sydney: Hale & Iremonger.

Riddel, R. 1994. Design. In R. Fisher and B. Crozier (eds) *The Queensland House: A Roof Over Our Heads.* The Queensland Museum, Brisbane.

Roderick, D.C. 2004. The origins of the elevated Queensland house. Doctorate of Philosophy, University of Queensland.

Sankalia, T. 2008. From facade to interstitial space: reframing San Francisco's Victorian residential architecture. In N. Alsayad (ed.) *The Cultural Politics of Dwellings: Traditional Dwellings and Settlement.* Centre for Environmnetal Design Research, University of California, Berkeley.

Skinner, P. 2011. An inside edge: urban infill potential within the suburban flood plain. *Association of Architecture schools of Australiasia (AASA) Proceedings, Deakin University.*

Smith, S. (1996). The tin camp: a study of contemporary aboriginal architecture in north-western New South Wales. Masters of Architecture Thesis, University of Queensland, St. Lucia.

Steele, W. 2011. Strategy-making for sustainability: an institutional learning approach to transformative planning practice. *Planning Theory and Research* 12(2): 205–221.

Steele, W. and Keys, C. 2015. Interstitial space and everyday housing practices. *Housing, Theory and Society* 32(1): 112–125.

Thompson, L. (2004). The indigenous living conditions problem: 'need', policy construction and potential for change. University of Queensland.

Tolia-Kelly, D.P. 2010. *Landscape, Race and Memory: Material Ecologies of Citizenship.* Ashgate, Farnham.

Wall, E. (2004). Memory and experience: the poetic spaces of the Queensand House. Bachelor of Architecture Thesis, University of Queensland, St. Lucia.

Watson, Donald. (1981). *The Queensland House: A Report into the Nature and Evolution of Significant Aspects of Domestic Architecture in Queensland.* Brisbane: National Trust of Queensland.

Yiftachel, O. 2012. Critical theory and 'grey space': mobilization of the colonized. *City*, 13(2–3): 240–255.

Zizek, S. 2006. *The Parallax View.* MIT Press, Cambridge, MA.

13 Burnt houses and the haunted home

Reconfiguring the ruin in Australia

Katrina Schlunke

Introduction

This chapter is an exploration of how we might think with the home that has become a ruin as a result of a natural disaster, in this case an Australian bushfire. It asks the reader to engage in a (re)thinking experiment rather than seeking confirmation of a simple critique of the meanings of home or the cultural cost of housing. It attempts to hold the idea of home and house in a kind of critical suspension through which experiences of loss, materiality, colonialism and the more-than-human pass through to re-enliven the ruin.

Natural disasters are organised within the alternating discourses of personal trauma and material destruction that evoke aspects of housing and senses of homelessness to make their points. This is particularly the case with bushfire in Australia, where the severity of the fire's effects will be shortened in news accounts to the number of houses and acres of land that have been burnt, followed by accounts of human suffering where specific individuals face the loss of their distinct homes (Muller 2011). This same truncation in how the destruction of fires is spoken of is reflected in the so-called recovery process where the technicalities of rebuilding homes and immediate prioritisation of bulldozing the remains sit alongside advice about surviving trauma, which often advocates going slowly and spending some time with the site of the former home. The recovery time for those people who lose their homes is spoken about as a seven-year process (Gordon 2011) while the timing for responses from human services, insurance agencies and demolition of burnt houses is days, weeks and months.

It is thus rare to find a contemporary ruin of a domestic home, given the exigencies of public safety and the demands for timely and public responses by some fire victims, local councils and national disaster teams. Using newspaper accounts and the personal experience of a home lost in bushfire, the ruin of the home is refigured in this chapter as a materialised connection between climate, the more-than-human and time whereby the usual opposition of the material and human is challenged.

Australia is a 'settler-colonial' nation that depended upon the notion of *terra nullius* (land belonging to nobody), to claim the right to reside upon and legally take over the Indigenous owned and occupied land (Rowse 1993). Although subsequent legal and political claims by Indigenous Australians have resulted in some acknowledgement of Indigenous land rights, there has been no recognition of continuing Indigenous sovereignty across the nation. The ruined home becomes a stopped moment where the non-Indigenous home is neither complete nor completely disappeared. In that stillness the 'ordinary' home can be seen more clearly as the colonial artefact it is and as a distinctly personal technology of white possession.

But the ruined home is also a call to continue homemaking, to clean up and to consider how and where the home came to be. Through the ruin, a new idea of home as a set of material effects that can neither be erased nor resurrected arises. The ruined home is instead a temporal marker. It indicates that a new arrangement of time and space has appeared. Not the expected one of developed house sitting on cleared land, where the modern defeats the natural, but a labile ruin of that expectation. In the place of the completed house is a set of material effects that arises from listening with and translating between things; firestorms, animals, assemblages of people and stuff that are violently thrown together and banally and incrementally sifted into each other through the drives of capital, changing weather and delicate webs of communication.

A home ruined by bushfire is in every sense a 'hot spot' that temporarily reconfigures the relations between the human and their homes. In the face of what the ruin of the domestic home might inspire are the policies of government and reactions of insurance companies to quickly demolish any ruin. This chapter, with its focus on embodied responses to the ruined home, suggests we should better imagine housing as something productively haunted by the ruin. That is, the ruin is a form of the home where we are able to see the environmental, political and colonial effects of those homes. Through the ruin we gain an appreciation of the finitude of colonial 'homes' while better recognising the Indigenous lands and ecological niches that are transformed through homemaking activities. As the manifestations of non-Indigenous people and so the marks in the landscape of an introduced species, houses become singular feral things but only seen as such through their ruin.

The imagined homes that organise this chapter are those that make up the ordinary streets of the small towns of Australia that are bordered by bushland or countryside. These homes will vary in size, shape and building material, but being produced through the same building codes and council compliance regulations will also have an order of commonness. They are often modest constructions of brick or weatherboard, given the cheapness of this land (at least compared to city prices) that attracts both those on low incomes and those wanting to be closer to the variably understood 'bush'. Given that the homes I am writing about have also been effected by bushfire, they are also different from those other ordinary homes that are in the streets of a city. Although they may look exactly the same, these home places have as part of their practised space (de Certeau 1984, 117) a relationship to wooded bushland, the usual site of homes effected by bushfire (although that is changing with the climate to include more cleared and open spaces as well).

Home, understood as a material figuration of 'bricks and mortar' co-joined with the emotional investment in those building materials and the artefacts that are held within it, is a modern and Western-centric vision that has remained pervasive while also being the site of multiple critiques and refinements. Morley provides an excellent overview of that work in 'Ideas of home' (2000, 16), which neatly unpacks both the historical shifts in how home and being 'at home' have changed over time and notes that home as sanctuary is 'far from being a universal experience'. He further notes that: 'The function and significance of home clearly varies with context' (Morley 2000, 29). For the purposes of this chapter there is still an organising principle that the house and so home is a separate special place which assumes the easy movement between the house that is built and becomes a key asset and the home that will be lived in. To pose a contingent separation, the home is the site of the emotional affect invested in it and the structure of the house, the frame, walls and roof are the material production of it. It is this separation that the ruin displaces to create a very different idea of home.

Ruin refigured

> Ruin is not a negative thing. First, it is obviously not a thing. One could write maybe with or following Benjamin, maybe against Benjamin, a short treatise on the love of ruins. What else is there to love, anyway? One cannot love a monument, a work of architecture, an institution as such except in an experience itself precarious in its fragility: it has not always been there, it will not always be there, it is finite. And for this very reason one loves it as mortal, through its birth and its death, through one's own birth and death, through the ghost or the silhouette of its ruin, one's own ruin, which it already is, therefore, or already prefigures. How can one love otherwise than in this finitude? Where else would the right to love, even the love of law come from?
>
> (Derrida 2002, 278; English translation only).

The ruin of a domestic home after bushfire is what remains before only the rubble of the 'cleared block' is left. To confront the ruin of one's home is, as Derrida suggests, to understand at last the love of 'home' that has now become precarious and fragile and so loveable precisely as a mortal and so finite kind of thing. It is in this instance of ruin that the bricks and mortar of the material home and the diurnal practices of making home lose their bounded opposition and become the affective and re-politicised event of the ruin.

Facing the ruin of a home is not a moment of visual undoing but of action. That action is usually referred to as 'fossicking', a process of sifting, searching and rummaging. Direct references and accumulated data about this ephemeral process are rare, and yet it is occasionally acknowledged by the press and incorporated into 'clean up' processes. In 2013 the *Blue Mountains Gazette* reported that one couple whose home was about to be demolished after bushfire 'had still been fossicking' (Curtin 2013). 'Still fossicking' meaning still touching and sifting and trying over and over to find a treasure and/or to create some sense of what matters and what doesn't and/or getting used to being homeless and/or saying goodbye touch by touch and/or…. Fossicking meaning the time spent both with and as the material of the house. A time in which the body experiences new ways of being and the 'new' home is felt through the body. In our part of the street where five houses were burnt down everyone came back at some stage, usually soon after the fire, to sift through, to check, to see. Many came back many times, long after there was any real expectation of finding any untouched objects but simply to sit with the ruin. In the Victorian fires of 2009, where fossicking was given a recognised role in 'disaster recovery', 14,000 'fossicker kits' were provided to fire victims that included face masks, gardening gloves, disposable overalls and waste collection bags (Karger *et al.* 2012, 40). In Victoria and other places, reports on fossicking emphasise its role in cleaning up and removing found objects, but so much more happens within the fossicked ruin.

The scale of what one can find is minute. Tiny fragments, wavelettes of ash, wholly extraordinary completeness in the shape of one intact cup now joined with lumps of glass, a cracked plate, a screw or a hammer rusted by heat. These small things are found amid the dislocation of the massive steel beams that once supported the roof which are now bent and fallen and the bricks and mortar crumbled and blackened. While the process of fossicking is wrapped in mourning it is also with every touch, with every faint effort to order the mess, a drawing of the ruin closer to oneself and closer to a homely effect. It is perhaps a ghostly re-enactment of the habits of domestic life. It may be an effort to enact the cycles of repeated care that created home via washing dishes, sweeping floors, picking up and putting down dusted objects and dirty clothes. In its insistence that some kind of homely habit is still possible, fossicking may offer a strange proof of Felski's claim about the everyday, home and gender that: 'repetition can signal resistance as well as enslavement' (1999, 21). The touching, sorting and caring is also an aesthetic practice that begins to group some things together as if also tentatively restoring the loss of loveliness. Derrida suggests of art that: 'A work is at once order and its ruin. And these weep for one and other' (2007, 122). Michael Duffy glosses this in his notes to suggest that the comprehension of these two orders at once is itself accessed through the tears of love that fall and prevent a simple sight of the painting/drawing and instead allows the vision proper (Duffy 2007).

In the literal ruin of the burnt home the movement between order and ruin is mimicked in the sifting and sorting of the dead objects and their ashes. Putting some things aside, releasing the ash back through the sieve while being covered oneself in that same ash as it drifts into hair, clothes, boots and skin. Marvelling at the forces that have come together – fire, glass, plastic, steel, electricity – that have made changes and produced awful, beautiful assemblages of stuff that produce, in turn, a stopping fascination. It is in this state that the home has become the labile, affecting work of art that is ruin but still suggests form.

The ruin and contemporary colonial settler society

Ginsberg writes that: 'When a viewer visits a ruin, the viewer is visited by the ruin' (2004, 171). Ginsberg has in mind the 'unified objects' (Gordillo 2014, 6) that are marketed as ruins. Places like the well-visited temples of Greece and Italy and/or at least the marked-out sites of more contemporary contributions to the heritage industry. But even in the intimate and more immediate ruin of the ordinary house the effects of the ruin are visited upon and within the viewer. It is the ruin that opens the house to the ruination of teleology. That is, the ruined domestic home shows us that as a nation and as managers of disaster, we can only think of clearing and re-building as quickly as possible. That the absence of a house reveals only a void, a new *terra nullius* asking, needing, to be populated. There is no response to the stopping or the stillness that the ruin also evokes which might act as an opening to the environment the home sits within and the history and politics of housing in a settler colonial context. Instead, as a spectral but serial assurance that 'civilised' progress might yet live, the ruin is denied again and again by the simple demand of citizen and government alike to rebuild. Again and again.

The home is literally opened to the environment and seen as ash mingling with the earth that the house once covered. The once properly housed and loved domestic artefacts become a litter of not always discernible materiality and in this state the thinking that depends upon a divide between nature and culture or between a surrounding environment and the home itself is itself ruined. In this instant the radical possibilities of the unbound home can not only be seen but also felt. In the space of the ruined home the material and the emotional are experienced as intertwined effects. Here the emotional home and the material house become lively stuff, doing stuff. As Bennett suggests of her 'enchanted materialism': 'Within this materialism, the world is figured as neither mechanistic nor teleological but rather as alive with movement and with a certain power of expression' (Bennett 2005, 447). This is a precise description of what it is to participate as human in the sifting home that has already been sifted by fire, wind and heat and previously by 'development' and capital to become house/home. And this site and stuff is lively not simply as a political imaginary or as a place managed emotionally through a narrowed down aesthetics but as a hotly effecting and affective space.

The difference between the two ideas of homes and houses, as emotional and material, are particularly stark in the face of their destruction in bushfire. 'Home' becomes the word associated with loss and feeling and 'house' or 'property' that which is associated with financial and organisational responses. These different meanings can be seen in the headlines and paragraphs from the reportage of a bushfire in Tasmania that mark the emotional space of endangered, saved or recovered homes. 'BUSHFIRE CRISIS: THE RECOVERY. Thanks. You saved my parents' home' (Rice 2015, 3) and 'New homes have sprung up quickly around bushfire ravaged Dunalley. But the feeling of being "at home" is taking a lot longer' (Richards 2014, 1). The home within these newspaper texts is unambiguously the space of particular emotional value that contains treasured artefacts and childhood memories. The heartfelt thanks to others for saving suggests the visceral nature of that feeling while the second headline gestures towards the difficulty of easily recreating that order of feeling in a new home. In between those headlines is the metamorphoses the home undergoes from existing cherished entity, to destroyed ruin, to being reorganised solely as a public hazard, a 'block' that needs clearing and the possible rebuilding and creation of the 'new' home.

To see a ruin of an ordinary home is to have the chance to see what wasn't actually there, what was occluded not just by the home itself but also by an unexamined belief in the idea of the home. Within the ruin one can see the ruination of a Western (and a very Australian) ideal of conquering space with location. We can see more clearly how Indigenous land and Indigenous knowledge that became the 'unknown' bush was held at bay by the home and the structure of the house. One order of teleological philosophy and bourgeois culture that promised European civilisation was lost with the Holocaust (Hofman 2005). But the remnant of that teleology, its shadowed remains, had a shape outside Europe in this 'settler colony' that went on and on filling up stolen space, often also in the name of 'civilisation'. 'Clearing' of the ruin is done in the name of the residents so those same residents can get on with rebuilding their new homes. The emphasis on re-building carries a sense of a return to the original – a possibility perhaps that through building alone, a restoration of all that has been lost might be restored and the home itself in some way renewed. It also means that the burnt block of land is a constant site of activity, never still enough, full enough by itself without a 'house' to evoke an original absence of Western housing, a pre-colonial time when the land was otherwise occupied – solely with Indigenous life.

What is left when even the rubble of the ruin is removed? What is a 'cleared block'? No destruction is ever that complete. There is always rubble and occasionally there are even small ruins of objects that are sufficiently intact, sufficiently auratic to recall the order of the whole home. But this is rare. The ordered destruction of the bulldozers that first crush the ruins and then remove the rubble to the 'waste facility' is done with such mechanical force that the top layer of soil is disturbed and blocks, far from being 'cleared' are scoured with the unmissable spoor of the caterpillar treads and lingering smell of diesel. The churned up ground creates ideal conditions for a whole range of introduced, invasive plant species that eventually follow the bulldozers onto these sites like mop-up troops. The block is both the building site to be and the site of ultimate loss, where even the trace is driven over as the ash is swept away. And yet it is not as it was before colonisation, when all 'nature' was also Indigenous culture. It is even less like land cared for within Indigenous thinking. It has a particular kind of silence that emanates from 'wild country' (Rose 2012) – the kind of country that has had Indigenous presence and story removed from it – and it has the silence of ordered numbness, being already ear-marked for a particular role like all the other shorn blocks in that same street.

In terms of houses and housing on a national scale, 'progress' is usually measured through the pace of development represented by the clearing of 'nature' to make way for initial roads and services which will in turn enable the construction of individual houses. In the case of the 'lost' home, however, the remains of the houses become themselves the site of the next 'development'. In this moment we catch a glimpse of the restless force that underlies 'home' building and re-building. It is a force where seemingly the new house consumes the remains of the old. This is a similar act but within a very different register to the deliberate erasure of the ruins of Palestinian villages by the Israeli military (Stoler 2013, 20). But knowing the remains of domestic homes are 'disappeared' in Israel provides a provocation to reconsider the remains of homes built over after an Australian bushfire.

The primary mode of 'legal' takeover of Indigenous land in Australia was through 'settlement'. And 'settlement' was most fully articulated through the development of systems of private ownership of which its most accepted and contemporary form is individual home ownership. This is a history and a present created through a colonial notion of 'progress' towards 'civilisation'. It is that same progressivist thinking that assumes what will come after a fire – the rebuilding of the home. Only rebuilding signals 'progress' and this is reflected in both the rhetorics of politicians and the everyday inquiries of concerned peoples who ask of the fire 'victim' as an expected and ordinary question: Are you rebuilding? Were you insured? At the same time as removing reminders of Indigenous sovereignty, the home is also imagined as a retreat from climate and a necessary protection from weather, which is only ever outside the home, kept at bay by the house. 'Rebuilding' suggests a heroic and active refusal of victimhood, but rebuilding also quickly covers up the remains that were themselves a possible reminder of a land without these kinds of 'bricks and mortar' homes. A land which was once entirely the home of Indigenous peoples.

The ruin as assemblage

The mixture of imaginings that are caught up in the idea of home becoming ruin and so home as hazard can best be exemplified by looking at a particular instance of homes effected by bushfire. What follows is a careful reading of extracts taken from a short article in the *Blue Mountains Gazette* following a bushfire in that area, combined with my own experiences of the bushfire that destroyed my home at that time. This combination of reportage, memoir and analysis reflects the ways in which the ruin itself is a 'combined act', an assemblage of state action, community concern, fire, nature and feeling.

Reporting on the bushfires that swept through the Blue Mountains in 2013, the then Premier of New South Wales (the state the fire occurred in), Barry O'Farrell described the effect of a bushfire that had recently occurred in the following way: 'The scale of the bush fire damage in the Springwood and Winmalee areas is immense, with more than 200 homes destroyed and another 120 properties damaged and the NSW Government is committed to supporting the area.' He went on to pledge his further support, stepping dexterously between home and hazard, in the following way: 'To assist, the NSW Government will work with insurers and fund the cleanup and removal of all bush fire debris, including the disposal of asbestos contaminated material from destroyed homes' (O'Farrell 2013, 1). It is that 'cleanup' that the rolling extracts below from the *Blue Mountains Gazette* refer to.

> By late last week, 77 blocks had been cleared. Crews from NRMA Insurance, Suncorp and Allianz are working seven days a week removing debris…. Blue Mountains MP, Roza Sage, who door-knocked some of the affected streets on Thursday with Finance and Services Minister Andrew Constance, was pleased at the progress made…
>
> 'The residents have made clear that they want this clearing to happen as quickly as possible and I cannot express how satisfying it is to see and hear the heavy machinery on the ground,' said Mrs. Sage…. 'The NSW Government, in partnership with insurers, is committed to ensuring the cleanup is done as quickly as possible, so residents can get on with rebuilding their homes.'
>
> …Mr. Constance and Mrs. Sage watched with Emma Parade resident Michael Magennis at Winmalee as the remains of his 10-year-old dream home were cleared. Mr. Magennis said the cleanup would allow his family to move on, although in a sense it was almost too soon as they had still been fossicking in the rubble to see if anything could be salvaged. 'My wife was here on Sunday night. She said it's the last time we can look at what we had', he said. They hope to rebuild on site but Mr. Magennis was finding it difficult to get quotes.
>
> …Mr. Koperberg *(Bushfire recovery coordinator)* said he was grateful to the owners of the Kemps Creek waste facility who had agreed to remain open last weekend to allow demolition crews to dump rubble.
>
> (Curtin 2013, 1)

The words of the conservative local politician wrest back from the 'natural disaster' and the temporal disorder of its aftermath, a return to teleological order – progress is being made. But 'progress' is now destruction. Progress is the number of blocks cleared and measured and the amounts of 'debris' removed. And yet the spectral 'home' is still invoked in the description of the politician's visit to the streets as 'door-knocking'. The door is still necessary to our imagining of the visit by both politicians as an amicable and appropriate action that was properly negotiated even if the material structures such as doors and windows that once allowed the vetting of visitors into a home are gone.

The expressed satisfaction of the local politician at hearing 'the heavy machinery on the ground' sets off other historical echoes. 'Heavy machinery' evokes both tanks rumbling into cities across the world and the ordinary labour of 'development'. The act of bulldozers destroying the remains of homes, under the aegis of 'cleaning up' are as much related to the images and experiences of human-wrought war as they are with organised responses to 'natural' disasters. In a world affected by climate change, this once seemingly clear division between man-made and 'natural' disaster no longer holds. The home that is now rubble being removed by that machinery may have had a role in the production of unsustainable, world-heating lifestyles, but in this moment that is unsayable. What must be reinstalled is the iteration of serialised cycles of development that rest upon rebuilding as a recovery of belief in housing and homes as they were before this 'natural' disaster. Before this interruption to housing as development.

The ambivalence of home as both site of new hope in the shape of a possible new one and the desire to stay with the remains, 'fossicking in the rubble' is poignantly at play in the extracts. The recognition that this clearing, this final destruction of the 'dream home' is not 'natural' but follows a timetable dictated by others is caught in Michael Magennis' suggestion that it was in a sense 'almost too soon', and his wife's words: 'She said it was the last time we could look at what we had.' The homeowners still recognise their home, maybe even their 'dream home' in its remains. They were still looking for something that 'could be salvaged'. Those whose intimate ruin it is understand the fruitfulness of the ruin and its possibilities. The work of homemaking does not cease simply because in every ordinary sense of the word, the home as shelter and protector has disappeared.

The economy that continues to drive the building and re-building of housing is alluded to in the Bushfire Recovery Coordinator's thanks to the 'waste facility' for staying open so that demolition crews could dump 'rubble'. Gordillo, in his compelling book *Rubble: The Afterlife of Destruction*, makes a distinction between the unified object of the ruin and the rubble that surrounds it that is disregarded (2014, 6). In the instance of the contemporary Western home and its politics of destruction, both ruin and rubble may help us reconsider what a home could be. Rubble is what can be seen and removed as quickly as possible after a natural disaster, but a ruin cannot. A ruin in this context cannot come into being except for that small window of time when 'fossicking' can occur.

For the home, the fire has rendered what was useful as useless and what was treasured is now trash. The political pressure for a politician to appear to be doing something is intense, and so these politicians are. The need for people to mourn and be with their homes is understood but not encouraged publicly. And as the rolling newspaper accounts suggest, all of that and more is in play as the home is refigured as rubbish.

When people return to a home ruined by fire or flood or other disaster, but where no human deaths have occurred, they will return to a governed site. When they return to the remains of their home that has been responded to with due diligence and state power working well in a basically democratic environment, they will return to a home surrounded by plastic hazard tape that warns the occupant away from their house. In different places the tape is red and white or yellow and black, and some has an inscription 'Do Not Cross' or 'Danger' repeated horizontally. Some particular trees, half burnt within the garden, will have their own hazard tape marking them for particular attention, specific destruction. The tape indicates that some preliminary risk assessment has been made and that the home has been recorded within several of the multiple systems of organisation that now arise from this site. This could include access to emergency relief because the home has been declared wholly uninhabitable, it could signal to any insurance assessor visiting the site that the house is considered by others to be totally destroyed and any claims are to now proceed along those lines, and most of all it signals to anyone passing that this is a dangerous site not to be entered for fear of one's own safety. The appearance of this tape also shouts (the tape is after all there to be very clearly seen) that this house has become one of many houses and other property sites that is now part of a 'crime scene' or an 'emergency response'. The tape calls the home into an association with the generic world of fictional crime and outlawed activity. The tape protects and castigates in the same moment – it produces the home as surveilled risk.

On the other side of the tape is the ruin. For the homeowner wishing to be with the surviving ruin the effect of the tape is more immediate and provocative. Few would obey its command utterly. But it creates in any re-association with the home an atmosphere of illegality rather than the sense of respected mourning that might accompany the visitation to other remains of loss. And it is not quiet. Within a scene that is otherwise almost silent, the tape catches every breeze, making an unsettling call that is somewhere between a thick plastic rubbish bag being scrunched and the unnerving vibration of paper-on-comb being badly played. It is the kind of noise that falls away for a moment only to reappear in some slightly new form as the breeze rises and falls or turns at a new angle in its attention seeking capriciousness.

This continuous recognition of the ways in which the home is partially produced through bureaucracy is one part of the heterogeneous flows that emerge in the finite space of this intimate ruin. In this space, 'disaster response' and the rewards of middle-class existence in the shape of insurance, emergency response teams and the policing of disaster space all become a sound that shifts at the will of the wind. The tape talks and talks. And it stages events. One story (which probably stands in for many other stories of the same ilk) is of an officer from the local council who comes through the tape to warn an owner that they should not be 'at home': that they might be endangering themselves and, worse, encouraging others, 'even children' to play in the blackened ruins of other homes along the street. The owner did not let the officer in the house or indeed into what was once the front garden bed – for what was now the boundary between house and outside? The owner made the officer back down because the ruin was still that powerful thing 'private property' and 'lost home' all at once, and its haunted effects still had social influence. That influence did not last long. As soon as an insurance claim was complete the demolition company hired by the insurers would erect tall cyclone fences around the ruin and it became a 'block to be cleared', inaccessible to everyone except those who came to do the demolishing.

In the instance of the bushfires in the Blue Mountains there is another colonial edge to this idea of hidden and refused ruin. The Blue Mountains National Park, only 130 kilometres from Sydney, is one of the oldest National Parks in Australia and is surrounded by the villages that run along the ridges of the so-called 'mountains'. The contemporary park's purpose includes 'protecting sites of Aboriginal cultural significance' (Blue Mountains National Park 2015a). National Parks have a complex relationship with Indigenous ownership and care of land. The 'preservation' philosophy that underscored the development of national parks across the world usually privileged 'environmental' values that resulted in the exclusion of Indigenous peoples from the land taken up into National Parks, and vetoes against Indigenous uses of the park for either hunting and/or cultural activities continue. That the park was once home to the Dharauk and Gundungurra peoples is well recognised (Blue Mountains National Park 2015b), but there is little active endorsement of the role the park might play in renewing and extending local Indigenous relations to place and their extended home. When fires break out in the national park, very brave specialist fire-fighters are sent in to attempt to preserve identified endangered ecosystems and key cultural sites, but these are never imagined as the protection of 'homes'.

The ruin as site of homemaking

As a ruin, the home is a sometimes-shameful place. There is nothing in the mess of bad housekeeping and broken down walls to hide within or behind. Shame, as Ferrell suggests, 'is being held in the power of another's gaze, like being caught in their headlights' (2004, 42). Without walls and ceilings the home as ruin is open to all, all can see. It is the ultimate failure of hospitality with no capacity to offer anything to anyone – no door to open in welcome nor a cup of anything, even the cup, to proffer. And it revolts in instances of horror. When a last frozen remain in the destroyed freezer finally thaws and comes alive with the smell of putrefaction and the bathroom in its mix of fallen-in ceilings and ashed water makes a sludge the colour of old corpses there is a rush of angry desire for it all to be gone. Utterly, utterly gone. Bring on those bulldozers, do that demolition. Stop this horror of seeing modernity fail us.

But the ruin also rises up as a promise of relief as multiple threads of memory entwine in strange new ways. One takes the shape of an automatic (wholly somatic?) walk. When the temperature drops and I feel cold I walk towards where the wardrobe used to be, to get a jumper, and am a metre or two along that route before I remember (or is it forget?) that neither cupboards nor clothes are there. There are constant little eruptions of corporeal memory that convince me and occasionally others that this space will still offer comfort and connection even as visually there is nothing that could do that work. It is not repression but a kind of confusion of boundaries between then and now, here and there, comfort and discomfort and this and that. The ruin makes of home a cultural ganglia where multiple experiences from different times and multiple assemblages from different matters all come into an over-excited order of connection that is created by the dissipating fire. Here, terms like house as structure and home as hearth no longer fulfil their promise to represent or suggest a meaning dependent upon an implied opposition. But this failure of sense and meaning also produces contingent grounds to figure a new kind of home/house.

When I fossicked in my home I found a scattered nest of tiny beads hidden in the heavy ashy dust of the former bedroom. They were a confused mix of the many necklaces we had owned. Not so very many but enough to 'dress up' for something special. As each bead emerged out of the ash and burnt scree of the sieve they recalled the strands around the neck, the 'getting ready' to go out, the closing of the clasp which were all one part of the final gestures that was 'leaving the house' to go out. These beads were a record of a culture, of the styled bourgeois acts of 'going to dinner', or a concert or celebrating something with someone special and going where beads went and took us. Those beads had blended with the leaves that had blown in and some of the glass ones were shattered from the inside out where the heat had swept through their tiny orifices and all were blackened although not so entirely that I didn't have a small hope they might be washed clean. The weeping between the perfectly performed bead and its ruin made of every one of them something to hold, to keep to oneself. In the ash they had become hidden treasures, unexpected and inspiring. I gathered them up and eventually I made a new necklace of those liquorice all-sorts colours, smudged with black, and they became another way to walk out of and yet be with the house/home.

Conclusion

Steven Wanta Jampijinpa Patrick writes about what happens to the locals when their land is taken and what has to be developed to get it back.

> This is what I call *Ngurra-kurlu. Ngurra-kurlu* is all about our place and sense of home. It consists of Family, Law, Land, Language, and Ceremony. Once we lose these five elements we become homeless people – people without the ability to understand our own home. We become

feral in our own land. We live in our home without really knowing how to look after it, and we run the risk of desecrating our home.

(Patrick 2015,123)

And what of those of us who were feral right from the start? We who understood an idea of home developed in a very different place? Can we come to love the ruin? The love of the ruin according to Duffy (2007) is the active, loving acceptance of the ruin. 'It is a ruining of oneself in front of a ruin, collapsing in front of it such that one respects its infinite inability to become anything other than a ruin, or restore oneself to a ruin. That is, it is an act that collapses again and again, infinitely, in front of its collapsing or ruining again and again. In other words, it is a respect for the ruin's self-deconstructing of itself by one's self-deconstructing of oneself, the iteration of one's traces in front of the other, tracing itself into the infinitely other. 'This is not a mutual relation, and it is an impossible relation, but it remains what happens in an ethical call to respect the ruination that occurs already in a text or drawing (or a person) composed and composing itself in traces' (Duffy 2007).

This seemingly histrionic re-reading of Derrida carries the feeling, the necessary performative release, of what it might be to displace the human from the centre of the home. Can we try instead to see through the ruin of the 'home' and the 'house', together with fire, wind, leaves, cups, political interventions, ash, love, bricks and mortar, horror and tears – the possibilities of re-envisaging home as affective assemblage? That is, a becoming in place? And could these dwellings, instead of another wall against the weather and the elements of climate, become very different, 'lively' things?

Could that feeling of 'ruination' be one part of thinking with the ruin to understand homes and houses not only as the major asset, the immoveable investments of love and bricks and mortar, but as a transient thing like others (including other feral things) which come into being in a place and change the people and the place in so doing (Muecke 2001). This new kind of 'home' is aware of the form it has been given through Western ideological frames and the ruin it is also, as colonial claim and climate changer. In its singular form this kind of home/house sees its postcolonial place alongside other distinctly introduced creatures that has to find its transient ecological niche with other things and other people on a vastly expanded material plane.

References

Bennett, J. 2005. The agency of assemblages and the North American blackout. *Public Culture* 17(3): 445–465.

Blue Mountains National Park 2015a. Introduction to. Retrieved on 20 October 2015 from www.nationalparks.nsw.gov.au/Blue-Mountains-National-Park.

Blue Mountains National Park 2015b. Ancient connections. Retrieved on 20 October 2015 from www.nationalparks.nsw.gov.au/visit-a-park/parks/Blue-Mountains-National-Park/Learn-More#EED33BEA9E3945CEA59351D841DF8093.

de Certeau, M. 1984. *The Practice of Everyday Life*. University of California Press, Berkeley.

Curtin, J. 2013. Fire clean-up starts to gain momentum. *Blue Mountains Gazette*. 17 December. Retrieved on 5 May 2015 from www.bluemountainsgazette.com.au/story/1974957/fire-clean-up-starts-to-gain-momentum.

Derrida, J. 2007. *Memoirs of the Blind: The Self-Portrait and Other Ruins*. University of Chicago Press, Chicago.

Derrida, J. 2002. Force of law. In J. Derrida and G. Anidjar, *Acts of Religion*. Routledge, London.

Duffy, Michael. 2007. Derrida, ruins, and love. Retrieved on 20 October 2015 from http://mikejohnduff.blogspot.com.au/2007/12/derrida-ruins-and-love.html.

Felski, R. 1999. The invention of everyday life. *New Formations* 39: 15–31.

Ferrall, R. 2004. *The Real Desire*. Indra Publishing, Briar Hill.

Ginsberg, R. 2004. *The Aesthetics of Ruins*. Rodopi, Amsterdam.

Gordillo, G. 2014. *Rubble: The Afterlife of Destruction*, Duke University Press, Durham, NC.

Gordon, R. 2011. The course of recovery after disaster. *CIMA Conference*, Melbourne, November. Unpublished Paper, retrieved on 20 October 2015 from www.cima.org.au/.../Rob-Gordon-The-Course-of-Recovery-after-Disaster.

Hofman, K. 2005. Poetry after Auschwitz: Adorno's dictum. *German Life and Letters* 58(2): 182–194.

Karger, H., Owen, J. and van de Graaf, S. 2012. Governance and disaster management: the governmental and community response to Hurricane Katrina and the Victorian fires. *Social Development Issues* 34(3): 40.

Morley, D. 2000. *Home Territories: Media, Mobility and Identity*. Routledge, London.

Muecke, S. 2001. Devastation. *UTS Review* 7(2): 123–129.

Muller, D. 2011. *Media Ethics and Disaster: Lessons from the Black Saturday Bushfires*. Melbourne University Press, Melbourne.

O'Farrell, B. 2013. Blue Mountains bush fires: funding provided for cleanup. Joint Media Release, 14 November. Retrieved on 6 May 2015 from www.ministerjustice.gov.au / Mediareleases/Pages/2013/Fourth%20Quarter/14November2013BlueMountains bushfiresFundingprovidedforcleanup.asx

Patrick, W.S. 2015. Pulya-ranyi, winds of change. *Cultural Studies Review* 21(1): 121–131. Retrieved on 3 October 2015 from http://epress.lib.uts.edu.au/journals/index.php/csrj/article/view/4420/4754

Rice, S. 2015. Headline. *The Advertiser*, Adelaide, 9 January, p. 3.

Richards, B. 2014. Headline. *The Mercury*, Hobart, 7 June, p. 1.

Rose, D.B. 2012. Why I don't speak of wilderness. *EarthSong Journal: Perspectives in Ecology, Spirituality and Education* 2(4): 9.

Rowse, T. 1993. *After Mabo: Interpreting Indigenous Traditions*. Melbourne University Press, Melbourne.

Stoler, A. (ed.) 2013. *Imperial Debris: On Ruins and Ruination*. Duke University Press, Durham, NC.

Thrown-togetherness, 2015
Exegesis

Andrew Gorman-Murray

The project

Thrown-togetherness draws conceptual precedents from both cultural geography and contemporary art theory and practice (Blunt and Dowling 2006; Perry 2013; Racz 2015). The project emerged as a response to this book's call for chapters, and was developed subsequently as a major work within a Master of Art at the UNSW College of Fine Arts. The driver from cultural geography is Doreen Massey's concept of space – including the home-space – as a thrown-together. Her propositions are that homes: are produced by relationships between the human and other-than-human, which are multi-scalar and stretch beyond the home-site; are spaces of multiplicity and heterogeneity in which diverse trajectories meet and co-exist; and are always in the process of being made and remade, never finished or closed. Home is therefore multiply lived and open to multivalent meanings and experiences.

Picking up this contention, the driver from contemporary art is George Kosuth's *One and Three Chairs* (1965), a conceptual artwork that juxtaposes a real chair, a full-size image of that chair, and a dictionary definition of chair. Thinking about this work in terms of sign systems, *One and Three Chairs* juxtaposes referent, signifier and signified in order to make explicit the multivalent readings of material objects, and to explore how 'reality' and experience lie at the junction of multifarious meanings. His work asks the viewer to consider the intertextuality and relationality of meaning.

Thrown-togetherness takes up *One and Three Chairs* as a conceptual starting point for speculating on the thrown-together home through visual language. The subject for this visual and material process is the artist-geographer's own home, Nicola Villa, a Victorian terrace located in Enmore in Sydney's inner western suburbs, originally built in 1887 but transformed continuously through the intersection of human and other-than-human presence. Acknowledging that meaning and experience are mediated spatially, *Thrown-togetherness* explores the home-space of Nicola Villa through three spatial frames: the external, the internal and the interstitial. Each series takes on further specific conceptual drivers and art-historical precedents to generate a visual language for the thrown-together home. The series and their objectives are described below; a selection of the colour images from each series has been reproduced in the book in black and white.

The external

The eight images in this series, five of which appear in the book, aim to document the external thrown-togetherness of Nicola Villa: the placement and form of the house-as-home in a context of temporal change and spatial flow. These images draw attention to the home – in this case, Nicola Villa – as a site within a world of both human and other-than-human movement, including the flows of transport systems (road/car, air/plane, rail/train), energy systems (wires, electricity), light, atmosphere, weather, diurnal change and transformations in building and infrastructure. This shows Nicola Villa not as a static site in a world of movement, but as a location that is itself in flux – through changes to the building structure, changes to the wires and poles connecting the house to grids and changes in the atmosphere, light and weather systems that ensconce the home.

The conceptual drivers for the external images are two-fold. Katsushika Hokusai's *36 Views of Mt. Fuji* (1826–1833) offers a visual language for placing the home in a context of material flows. Hokusai's images visually frame Mt. Fuji through various natural and cultural elements – barrels, bridges, clouds, waves, kites, scaffolds and trees, and with foreground action of people – in order to show how the significance of the mountain reaches across the spheres and sites of life in nineteenth-century Japan. The roof of Nicola Villa has visual semblance to Mt. Fuji, and is the iconic element of the subject (house) in the images. The images therefore place Nicola Villa and its distinctive roof in frames of wires, poles, signs, roads, clouds, surrounding buildings, planes and streetlights.

The other conceptual driver is the work of photographer Gregory Credwson, particularly his night-time series of small-town and suburban America, notably *Twilight* (1998–2002). The use of light in his compositions transforms the mundane buildings of small towns and suburbia into surreal narratives of place. His work shows the critical way in which light and temporality mediate the production and experience of home. I followed his precedent in creating the night-time and twilight (and daytime) external images of Nicola Villa.

The internal

The eight images in the internal series, four of which are reproduced in the book, document the behavioural traces of domestic life – traces that are in constant motion, ephemeral, never fully within the grasp of human agency. They are traces that include human and other-than-human entities within the domestic sphere: companion animals, objects, water and energy systems, atmosphere and light. These messy behavioural traces challenge the 'pristine' and 'private' domestic environment represented in interior design magazines. These images offer a visual language that captures Nicola Villa as a relational, processual and multiply lived space. The final images represent various interior spaces throughout the house at different times of the day. In terms of human presence, they document everyday domestic activities that are always performed in a context of both human and other-than-human relations.

The composition departs from and questions stereotypical domestic interiors in two further ways: they are aerial shots, taken from above the rooms; and they include the shocking presence of feet on a ladder. The latter inclusion, on the one hand, alludes to the processual nature of both homemaking and photography. On the other hand, bare feet are symbolic in human society and culture, denoting a range of contradictory meanings: privacy, servile status, proximity to nature, freedom from constraint, innocence and humility. Bare feet can be a sign of respect when entering holy ground: 'Remove the sandals from your feet, for the place where you are standing is holy ground' (Exodus 3:5 NRSV). That the feet (and presumably the person) are atop a ladder also suggests the precarity and dis-ease of home, not only sanctity and privacy. I do not fix the meaning of bare feet in these images, but leave the possibilities open.

The visual and conceptual precedents for these images include the work of photographers Stephen Shore (*Uncommon Places*, 1982) and Simryn Gill (*Dalam*, 2001), both of which aim to capture and document the messy and diverse behavioural traces of everyday domestic life. They seek to make the mundane and overlooked into something of interest and concern through composition and attention to detail. I have sought the same – capturing unexpected detail. To generate a provocative and compelling visual language for these details, I have been inspired by the precedent of Izabela Pluta's *Front Yards* (2004), which documents aerial views of residential front yards in Kalgoorlie. The unexpected perspective prompts the viewer to look at these mundane domestic spaces from a fresh angle and notice details typically overlooked. The composition changes how we might think about place and inhabitation.

The interstitial

The eight images in the interstitial series, three of which appear in the book, explicitly explore other-than-human presence within the domestic sphere – specifically the way other species make home, and unmake human meanings of home, within domestic space. These images comprise triptychs, each juxtaposing 'daddy long-leg spiders' within a domestic room with a slug nesting on and eating a domestic bill, and a spider-, slug- or cockroach-eye view of the interior space. The interior view includes further human and other-than-human presences and activities within the space, suggesting that the relational, multiple and processual experience of home is not just experienced from a human-centred perspective, but by other-than-human inhabitants too. Spiders use materials erected by humans to make home; slugs unmake the objects that come into the home, in this case domestic bills. Who is this home for?

The conceptual driver for this series is the work of Catherine Chalmers, who anthropomorphises and photographs domestic insects in *American Cockroach* (2004) and *Houseflies* (1999). Visually interrogating how insects make home in 'human homes' prompts my exploration of spiders and slugs at home, and their relations with other inhabitants. However, I do not want to anthropomorphise them, but rather document their own capacity and agency as other-than-human beings. In this regard I follow the precedent of Simryn Gill's *Wormholes* (2009), which documents the effect of worms and insects on abandoned buildings and objects, exploring other-than-human agency through visual language.

I sought to develop my own visual language in order to speculate on the multiply lived home, or the cohabitation of human and other-than-human. To do this I developed triptychs, which also allowed me to experiment with scale and composition, to give the other-than-human a spatial presence in the home, as species that interact with human activities and with companion animals – in other words, to visualise inter-species relations within the domestic sphere. The other-than-human presence is part of the making and unmaking of home.

Openings

Thrown-togetherness offers a creative, visual and material approach to the multiple agencies, co-existences and interactions within, and that transform, everyday domestic geographies. To do this, I have taken on the role of an artist-geographer working at the intersection of cultural geography and contemporary art. This is a productive site at and in which to work. It enables the use of visual language to speculate on lived geographies, and to wonder over and over again through active, contemplative viewing. This speculative visual language seeks to open up possibilities for thinking through the unbounded home.

References

Blunt, A. and Dowling, D. 2006. *Home*. Routledge, London.
Massey, D. 2005. *For Space*. Sage, London.
Perry, G. 2013. *Playing at Home: The House in Contemporary Art*. Reaktion, London.
Racz, I. 2015. *Art and the Home: Comfort, Alienation and the Everyday*. Tauris, London.

Index

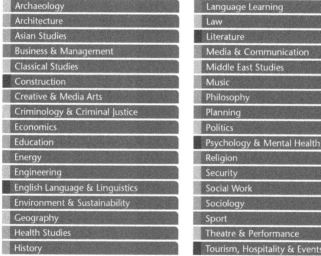